AF334455

Tobias Grosche

Computational Intelligence in Integrated Airline Scheduling

Studies in Computational Intelligence, Volume 173

Editor-in-Chief

Prof. Janusz Kacprzyk
Systems Research Institute
Polish Academy of Sciences
ul. Newelska 6
01-447 Warsaw
Poland
E-mail: kacprzyk@ibspan.waw.pl

Further volumes of this series can be found on our
homepage: springer.com

Vol. 152. Jarosław Stepaniuk
*Rough – Granular Computing in Knowledge Discovery and
Data Mining*, 2008
ISBN 978-3-540-70800-1

Vol. 153. Carlos Cotta and Jano van Hemert (Eds.)
*Recent Advances in Evolutionary Computation for
Combinatorial Optimization*, 2008
ISBN 978-3-540-70806-3

Vol. 154. Oscar Castillo, Patricia Melin, Janusz Kacprzyk and
Witold Pedrycz (Eds.)
Soft Computing for Hybrid Intelligent Systems, 2008
ISBN 978-3-540-70811-7

Vol. 155. Hamid R. Tizhoosh and M. Ventresca (Eds.)
Oppositional Concepts in Computational Intelligence, 2008
ISBN 978-3-540-70826-1

Vol. 156. Dawn E. Holmes and Lakhmi C. Jain (Eds.)
Innovations in Bayesian Networks, 2008
ISBN 978-3-540-85065-6

Vol. 157. Ying-ping Chen and Meng-Hiot Lim (Eds.)
Linkage in Evolutionary Computation, 2008
ISBN 978-3-540-85067-0

Vol. 158. Marina Gavrilova (Ed.)
*Generalized Voronoi Diagram: A Geometry-Based Approach
to Computational Intelligence*, 2009
ISBN 978-3-540-85125-7

Vol. 159. Dimitri Plemenos and Georgios Miaoulis (Eds.)
Artificial Intelligence Techniques for Computer Graphics, 2009
ISBN 978-3-540-85127-1

Vol. 160. P. Rajasekaran and Vasantha Kalyani David
Pattern Recognition using Neural and Functional Networks,
2009
ISBN 978-3-540-85129-5

Vol. 161. Francisco Baptista Pereira and Jorge Tavares (Eds.)
Bio-inspired Algorithms for the Vehicle Routing Problem, 2009
ISBN 978-3-540-85151-6

Vol. 162. Costin Badica, Giuseppe Mangioni,
Vincenza Carchiolo and Dumitru Dan Burdescu (Eds.)
Intelligent Distributed Computing, Systems and Applications,
2008
ISBN 978-3-540-85256-8

Vol. 163. Pawel Delimata, Mikhail Ju. Moshkov,
Andrzej Skowron and Zbigniew Suraj
Inhibitory Rules in Data Analysis, 2009
ISBN 978-3-540-85637-5

Vol. 164. Nadia Nedjah, Luiza de Macedo Mourelle,
Janusz Kacprzyk, Felipe M.G. França
and Alberto Ferreira de Souza (Eds.)
Intelligent Text Categorization and Clustering, 2009
ISBN 978-3-540-85643-6

Vol. 165. Djamel A. Zighed, Shusaku Tsumoto,
Zbigniew W. Ras and Hakim Hacid (Eds.)
Mining Complex Data, 2009
ISBN 978-3-540-88066-0

Vol. 166. Constantinos Koutsojannis and Spiros Sirmakessis
(Eds.)
Tools and Applications with Artificial Intelligence, 2009
ISBN 978-3-540-88068-4

Vol. 167. Ngoc Thanh Nguyen and Lakhmi C. Jain (Eds.)
Intelligent Agents in the Evolution of Web and Applications,
2009
ISBN 978 3 540-88070-7

Vol. 168. Andreas Tolk and Lakhmi C. Jain (Eds.)
*Complex Systems in Knowledge-based Environments: Theory,
Models and Applications*, 2009
ISBN 978-3-540-88074-5

Vol. 169. Nadia Nedjah, Luiza de Macedo Mourelle and
Janusz Kacprzyk (Eds.)
Innovative Applications in Data Mining, 2009
ISBN 978-3-540-88044-8

Vol. 170. Lakhmi C. Jain and Ngoc Thanh Nguyen (Eds.)
*Knowledge Processing and Decision Making in Agent-Based
Systems*, 2009
ISBN 978-3-540-88048-6

Vol. 171. Chi-Keong Goh, Yew-Soon Ong and Kay Chen Tan
(Eds.)
Multi-Objective Memetic Algorithms, 2009
ISBN 978-3-540-88050-9

Vol. 172. I-Hsien Ting and Hui-Ju Wu (Eds.)
Web Mining Applications in E-Commerce and E-Services,
2009
ISBN 978-3-540-88080-6

Vol. 173. Tobias Grosche
Computational Intelligence in Integrated Airline Scheduling,
2009
ISBN 978-3-540-89886-3

Tobias Grosche

Computational Intelligence in Integrated Airline Scheduling

 Springer

Tobias Grosche
Meerfeldstr. 52
68163 Mannheim
Germany
E-Mail: tobias.grosche@gmail.com

ISBN 978-3-540-89886-3   e-ISBN 978-3-540-89887-0

DOI 10.1007/978-3-540-89887-0

Studies in Computational Intelligence   ISSN 1860949X

Library of Congress Control Number: 2008941002

Typeset & Cover Design: Scientific Publishing Services Pvt. Ltd., Chennai, India.

Printed in acid-free paper

9 8 7 6 5 4 3 2 1

springer.com

Acknowledgements

This study is based on my Ph.D. thesis completed at the Department of Business Administration and Information Systems I at the University of Mannheim. Doing this research and finishing this thesis would not have been possible for me without the support and encouragement of many people. Their effort not only contributed considerably to this work but also their company made the time of doing it worthwhile and enjoyable. At this point, I would like to take the opportunity to express my gratitude to some of them as representatives for many.

Foremost, I have to thank Prof. Dr. Armin Heinzl for giving me the opportunity to complete this thesis at his department. Having a passion for aviation it was to my delight that he gave me the freedom to do research in the field of airline operations research.

I am especially grateful to Prof. Dr. Franz Rothlauf who had a significant impact on both, my research work and my commitment to it. Since my first steps in research until the final technical preparation of the manuscript he always provided a steady and reliable support and encouragement that was essential for me and my work. And as a good friend he also helped me to manage non-research related ups and downs.

My colleagues and friends at our department provided an atmosphere that was much fun to work in. A large portion of my motivation depended on them, and they always were of great help to share and to overcome daily troubles. I am grateful to every one of them for the great time we spent together at the university and beyond.

In particular, I want to thank Thomas Butter. He together with Franz Rothlauf provided patient and skilled assistance with programming and implementation issues which was of great relief to me. I am grateful to Kerstin and Pauline Strowa for their very thorough and proficient proof-reading.

At last, though not least, I would like to thank my friends and family. My parents supported my educational and academic career as much as they could. They always encouraged me in my way and in pursuing my goals from the very beginning. Without them, their support, and their belief in me, I would never have gotten into the situation to be able to write this thesis.

Contents

List of Figures

List of Tables

Acronyms

3-MET	Three-day maintenance Euler tour
AEA	Association of European Airlines
ATA	Air Transportation Association of America
ATC	air traffic control
BB	building block
BTS	Bureau of Transportation Statistics
CRS	computer reservations system
CS	calibration set
DSS	decision support system
EA	evolutionary algorithm
EIS	custom model for the estimation of itinerary shares
ES	evolution strategy
FAA	Federal Aviation Administration
FIFO	first-in-first-out
GA	genetic algorithm
GDP	gross domestic product
GDS	global distribution system
GP	genetic programming
IATA	International Air Transport Association
ICAO	International Civil Aviation Organization
IP	integer programming
LCC	low-cost carrier
LIFO	last-in-first-out
LL	log-likelihood
LOF	line-of-flying
LP	linear programming
MIDT	Market Information Data Tapes
MLE	maximum likelihood estimator
MNL	multinomial logit (model)
MSE	mean squared error
NFL	No-Free-Lunch Theorem

NUTS	Nomenclature des unités territoriales statistiques
OAG	Official Airline Guide
OR	operations research
O&D	origin and destination
PAX	passenger(s)
RPK	revenue passenger kilometer
RPM	revenue passenger mile
SA	simulated annealing
SCP	set covering problem
SD	standard deviation
SLF	seat load factor
SPP	set partitioning problem
TA	threshold accepting
UTC	Universal Time Coordinated
VS	validation set
rGA	selecto-recombinative genetic algorithm

Chapter 1
Introduction

Abstract. An airline schedule represents the central planning element of each airline. In general, the objective of airline schedule optimization is to find the airline schedule that maximizes operating profit. This planning task is not only the most important but also the most complex task an airline is confronted with. Until now, this task is performed by dividing the overall planning problem into smaller and less complex subproblems that are solved separately in a sequence. However, this procedure is only of minor capability to deal with interdependencies between the subproblems, resulting in less profitable schedules than those being possible with an approach solving the airline schedule optimization problem in one step. In this work, two planning approaches for integrated airline scheduling are presented. One approach follows the traditional sequential approach: existing models from literature for individual subproblems are implemented and enhanced in an overall iterative routine allowing to construct airline schedules from scratch. The other planning appraoch represents a truly simultaneous airline scheduling: using metaheuristics, airline schedules are processed and optimized at once without a seperation into different optimization steps for its subproblems.

1.1 Overview

Airline operations are essential for economic development and growth. By offering both freight transportation and passenger travel, the airline industry supports an economy that is based on principles of human cooperation and of distributed production and consumption of goods and services in a competitive environment. In 2005, for the first time the number of passengers of the world's scheduled airlines exceeded two billions (ICAO, 2006). Between 1995 and 2005, the number of revenue passenger kilometers (RPK) of scheduled airline traffic grew with an average annual rate of 5.2% (see Fig. 1.1). According to the forecasts of the two major aircraft manufacturers Airbus and Boeing, this trend is expected to continue with an average increase of 4.8% to 4.9% RPK per year until 2025 (ICAO, 2006; Airbus, 2006; Boeing, 2006).

T. Grosche: Computational Intel. in Integrated Airline Scheduling, SCI 173, pp. 1–5.
springerlink.com

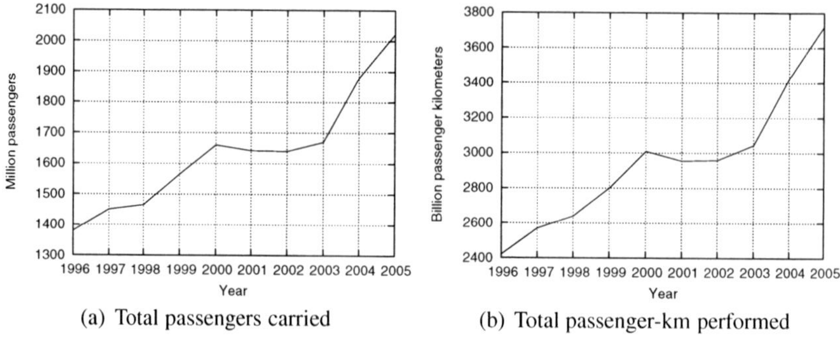

(a) Total passengers carried   (b) Total passenger-km performed

Fig. 1.1 Passenger traffic increase 1996-2005 (Source: ICAO (2006))

However, despite positive market trends, the airlines' profit margin is considerably small and shows the lowest performance of any of the individual sectors in the air transport sector (Doganis, 2004). As the integral components of an airline are personnel and aircraft, representing fixed and most expensive costs (Butchers et al., 2001), the annual net profit or loss of the airlines mainly depend on passenger demand. This demand is highly correlated with the overall economic growth. As a consequence, the airlines' profits usually follow a cyclical trend with economic downturns and booms (see Fig. 1.2).[1]

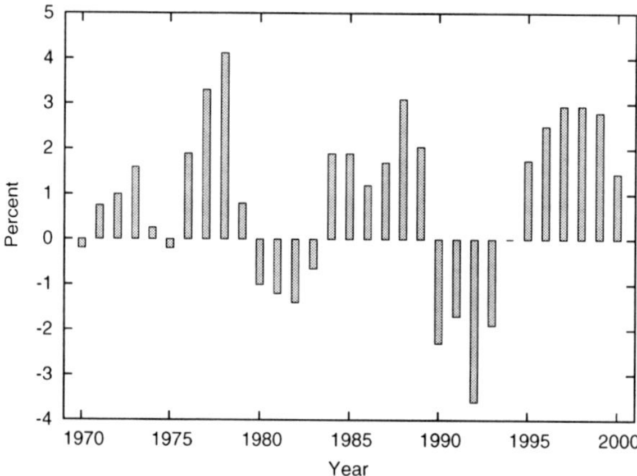

Fig. 1.2 Annual net profit or loss as a percentage of total revenue of ICAO member airlines, 1970-2000 (Source: Doganis (2004))

One reason for the airlines' marginal profitability might be the increase of competition and rivalry after the liberalization of the airline markets in the USA in 1978 and in Europe in 1993. The average revenue per seat kilometer has declined to 50% in the

---

[1] In 2005, the worlds airlines had a combined operating profit of 1% of operating revenues (ICAO, 2006).

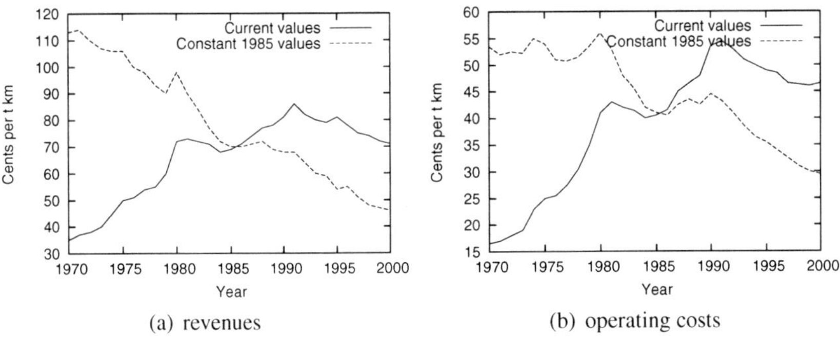

Fig. 1.3 Trend in unit revenues and operating costs, ICAO scheduled airlines, 1970-2000 (Source: Doganis (2004))

last 20 years (Costaguta & Resiak, 2002; ATA, 2003b). In addition, the emergence of low-cost carriers (LCC) and improved information systems allow passengers to find low air ticket prices more easily (Button, 2005), further putting weight on operating yields. On the other hand, as Fig. 1.3 illustrates, airlines were able to compensate the decreased yields by reducing operating costs. The main drivers for this reduction were and still are the introduction of more modern and efficient aircraft, lean operating structures and the computerization in airline planning processes.

According to Doganis (2004), "there is no simple explanation of the apparent contradiction between the airline industry's rapid growth and its marginal and cyclical profitability. But, for the individual airline, overcoming this contradiction means matching supply and demand for its service in a way which is both efficient and profitable. This is the essence of airline management and planning. It is about matching the supply of air services, which management can largely control, with the demand for such services, on which management has much less influence." The result of this planning task is the airline's schedule: an optimal schedule represents the most efficient and effective deployment of an airline's resources while best satisfying potential passengers' demand (Etschmaier & Mathaisel, 1985). Being the central element within an airline's corporate planning system, it affects almost every operational decision and has the largest impact on profitability (Teodorovic, 1988; Suhl, 1995; Barnhart & Talluri, 1997; Barnhart et al., 2003).

From a marketing-perspective, the airline schedule has to offer flights that passengers demand. From a production-perspective, a cost-minimizing allocation of aircraft and crews has to be specified to operate the flights. These two perspectives are reflected by the flight schedule and the resource assignment as the two essential parts of an airline schedule. The flight schedule is presented to potential passengers and is the airline's primary product, having the most influence on a passenger's choice of an airline (Gopalan & Talluri, 1998b). It determines the flights and the route network of an airline. Nonstop flights as well as connection flights including their departure and arrival times at the airports and their days of operation are published to the customers (see Fig. 1.4 for example). The

Dep Ab	Arr An	Flight Flug	Type Typ	Freq Freq	St ST	Meal ML	Dur Dur		Dep Ab	Arr An	Flight Flug	Type Typ	Freq Freq	St ST	Meal ML	Dur Dur
Olbia							**(OLB)**		**Osaka**							**(KIX)**
München/Munich (MUC)							**527mi**		**Frankfurt (FRA)**							**5,760mi**
15.15	17.00	**LH2641✦**	AR8	7	0	S	1:45		09.55	15.05	**LH741**	346		0	LL	12:10
16.55	18.35	**LH2641✦**	CR7	6	0	S	1:40		**Fukuoka (FUK)**							**280mi**
Omaha							**(OMA)**		10.15	11.25	**LH9792◊**	735		0		1:10
Chicago (ORD)							**415mi**		Above Disc. 5/31							
07.44	09.05	**LH9337✦**	752	6	0		1:21		12.40	13.50	**LH9792◊**	320		0		1:10
Above Disc. 4/21									Above Eff. 6/1							
08.04	09.25	**LH9337✦**	752	X6	0		1:21		**Kagoshima (KOJ)**							**318mi**
Above Disc. 4/23									14.50	16.00	**LH9894◊**	320		0		1:10
08.31	09.52	**LH5507✦**	CRJ		0		1:21		**Okinawa Naha JP (OKA)**							**727mi**
Above Eff. 4/24									10.00	12.00	**LH9890◊**	737		0		2:00
10.25	11.58	**LH9107✦**	735		0		1:33		Above Disc. 5/31, Exc. 5/10 - 5/16							
Above Eff. 6/7									14.05	16.05	**LH9890◊**	772		0		2:00
10.41	12.11	**LH9107✦**	319		0		1:30		Above Eff. 6/1							
Above 4/24 - 6/6, Exc. 4/25									**Sapporo (CTS)**							**673mi**
10.47	12.19	**LH5507✦**	CRJ	7	0		1:32		11.25	13.20	**LH9794◊**	320		0		1:55
Above Disc. 4/22									Above Eff. 6/1							
10.57	12.29	**LH9107✦**	733	X7	0		1:32		12.50	14.45	**LH9794◊**	320		0		1:55
Above Disc. 4/25, Exc. 4/24									Above Disc. 5/31							

Fig. 1.4 Lufthansa flight schedule - excerpt (Source: Lufthansa (2007))

passenger's choice of a flight is also influenced by the air fare. In order to be able to offer reasonable air fares and to minimize operating costs, it is necessary to distribute the airline's resources (aircraft and crews) in an effective and efficient way. Moreover, the airline schedule affects almost every operational decision, and on average 75% of the overall costs of an airline are directly related to the schedule. Thus, given an airline schedule, a significant portion of costs and revenues is fixed (Etschmaier & Mathaisel, 1985; Suhl, 1995; Seristö & Vepsalainen, 1997; Barnhart & Talluri, 1997; Langerman & Ehlers, 1997; ATA, 2002; ATA, 2003a).

In general, the objective of airline schedule optimization is to find the airline schedule that maximizes operating profit. This planning task is not only the most important but also the most complex task an airline is confronted with. Factors such as demands in various markets, competition, and available resources have to be considered simultaneously to achieve optimal solutions (Gopalan & Talluri, 1998a; Taneja, 2002; Barnhart et al., 2003). Because of the complexity of this problem, an airline schedule is usually constructed in several steps or stages emerging from a decomposition of the overall airline scheduling problem into smaller subproblems. The output of one subproblem represents the input for the next subproblem. Because of the reduced complexity, the subproblems can be solved more easily. In addition, if a subproblem is well structured, it can be solved using automated routines and optimization algorithms. Much research has been carried out on different models and algorithms for these subproblems. On the other hand, many decisions are conducted manually by human experts. They are supported by decision support systems (DSS) which help to assess the experts' proposals, detect any violations of restrictions or rules (Grandeau et al., 1998), or implement optimization algorithms for the solution of selected subproblems. DSS are also necessary to administer the complete scheduling process. Their support is necessary, because many feedback loops and iterations between the subproblems have to be implemented in order to consider existing interdependencies between the subproblems and their variables to complete the scheduling process.

1.2 Objective

Many sophisticated planning tools and optimization algorithms for the subproblems have been developed. However, it remains questionable if the decomposition of the overall airline scheduling problem reduces the quality of the resulting schedules compared to schedules that would result from an integrated airline scheduling approach. All researchers agree that the integration of the subproblems would lead to better schedules. In fact, as will be presented later, a trend towards publications of integrated models can be observed. But on the other hand, an integrated model of the complete airline scheduling problem is believed to be computational intractable and even its formulation seems impossible.

The objective of this study is to fill a large portion of the gap between theory's (or researchers') ultimate goal of a fully integrated airline scheduling approach and the status quo of sequential airline schedule optimization. Two approaches to integrated airline scheduling for scheduled passenger airlines are presented and evaluated. The first approach implements an itcrated sequential planning paradigm in an integrated procedure, whereas the second approach represents a truly simultaneous optimization approach without decomposition of the overall problem. Both approaches can be used to construct and optimize airline schedules from scratch. Given some basic parameters and planning scenarios, each method produces a feasible schedule that promises high operating profit. Compared to published contributions, airline operations are represented at a higher level of detail, simplifying assumptions are reduced, and additional practical restrictions are included.

1.3 Structure

This study is structured as follows. In the next chapter, an overview of the airline scheduling process is given. It presents the decomposition of the overall problem, the resulting subproblems, and optimization models to solve these subproblems. In addition, approaches for the integration of selected subproblems are described. Based on this literature review, future challenges are identified that motivate this study. In Chapter 3, metaheuristics and their foundations are introduced, because the simultaneous airline scheduling approach is based upon these optimization techniques. Chapter 4 represents the main chapter of this study. Within this chapter, Sect. 4.2 presents a schedule evaluation procedure that is used by the two integrated scheduling approaches. The sequential airline scheduling approach is described in Sect. 4.3, the simultaneous approach in Sect. 4.4. Both approaches are calibrated and analyzed, before a comparison of both approaches is conducted in Sect. 4.5. Finally, Chapter 5 includes a summary, conclusion, and an outlook on future work. In the appendix, a glossary is given containing aviation-specific terms used throughout this study. Furthermore, it includes information on the aircraft data and scenarios of the conducted experiments as well as their detailed results.

Chapter 2
Airline Scheduling Process

Abstract. In this chapter, the objectives, inputs, and constraints of the subproblems of the airline scheduling problem are presented together with solution models from existing literature to solve them. Because optimal solutions of the airline scheduling problem can only be realized if all relevant variables, their interdependencies, and restrictions are combined in one model of considerable detail, a trend towards an integrated airline scheduling model can be recognized in recent publications. Advances in optimization theory and computer hardware also led to the consideration of more realistic models, as a problem could be formulated in more detail with a higher number of practical requirements, with less simplifications, and for more realistic problem sizes. Although much effort has been undertaken and many sophisticated models for the airline scheduling problem have been developed, many challenges still remain. In particular, stochastic elements should be incorporated in the scheduling process to increase the robustness of the resulting schedules, optimization methods should be used in a larger number of subproblems that are still solved manually, airline operations have to be represented at a higher level of detail (reducing simplifying assumptions and including practical restrictions), and, finally, boundaries between the subproblems in the planning process should be further relaxed towards an overall integrated approach.

2.1 Introduction

2.1.1 Airline Scheduling

The airline schedule represents the central planning element of each airline. It is the instrument to match the available resources to the given demand. The airline schedule includes the flights of the airline including their departure and arrival airports and times, their days of operation, and the assigned fleet types. In addition, from a production perspective, it includes the assignment of specific aircraft and their related maintenance schedule and the assignment of cockpit and cabin crews. Every operational decision depends on the airline schedule, thus, once an airline schedule

T. Grosche: Computational Intel. in Integrated Airline Scheduling, SCI 173, pp. 7–46.
springerlink.com © Springer-Verlag Berlin Heidelberg 2009

is constructed, a large portion of the airline's costs is determined. In addition, because the schedule influences the number of passengers the airline will transport, it also affects the revenues the airline will gain.

The goal of airline scheduling is to create an airline schedule that is optimal in regard to a given objective, usually operating profit. This problem is usually solved in a structured process in which all parts of the airline participate. There are two perspectives on this process (Etschmaier & Mathaisel, 1985; Antes, 1998). One perspective focuses on the time axis: the planning process is divided into strategic, tactical and operational planning. During the strategic planning phase, long-term decisions are made and the framework for the subsequent decisions is constructed. Tactical decisions focus on specific flights and create a plan of action for the airline's operations. In this phase, most of the airline schedule's elements are selected. Finally, the operational phase includes adjustments of the schedule due to changes in demand or supply or any unforeseeable disturbances. The other perspective on airline scheduling focuses on the internal planning approach in an airline and its planning department. It describes airline scheduling as an iterating cycle of schedule construction and evaluation. The airline schedule or its elements are constantly modified and optimized until a satisfactory schedule is found or the planning time is over. Both perspectives describe the same planning process, they are not exclusive but used in conjunction by the airlines. For example, to create the schedule in the tactical phase, the planning departments of an airline usually propose different schedule drafts and incrementally improve these variants and their elements through cycling between evaluation and modification. Then, once the schedule is published, continuous refinements and modifications take place until the day of operation to encounter any disturbances or to include any changes or new and more detailed information of demand or supply.

Although there are different perspectives on the airline scheduling process, researchers agree that creating an airline schedule represents one of the most complex challenges an airline has to face. This complexity derives from the number of decision variables that have to be selected to create an airline schedule, the given (internal and external) data and restrictions that have to be taken into account, and the heterogeneity, functional relationship and interdependencies between all these factors (see Fig. 2.1). Optimal solutions to a given problem can only be obtained if all these elements are taken into account simultaneously in considerable detail in one optimization model and if exact algorithms are applied that guarantee the optimal solution is found. Because of the complexity of the airline scheduling problem, such a model has not been solved or even formulated (Langerman & Ehlers, 1997; Barnhart et al., 2003). With state-of-the-art solution algorithms and computer hardware, a model that solves the complete airline scheduling problem in one step is believed to be computational intractable (Hane et al., 1995; Suhl, 1995; Desaulniers et al., 1997; Antes et al., 1998; Barnhart et al., 1998; Klabjan et al., 2002; Barnhart et al., 2003; Lohatepanont & Barnhart, 2004).

The traditional approach to solve the airline scheduling problem is to decompose this problem into subproblems and to solve these subproblems in a sequential order (Suhl, 1995; Mathaisel, 1997; Grandeau et al., 1998; Barnhart et al.,

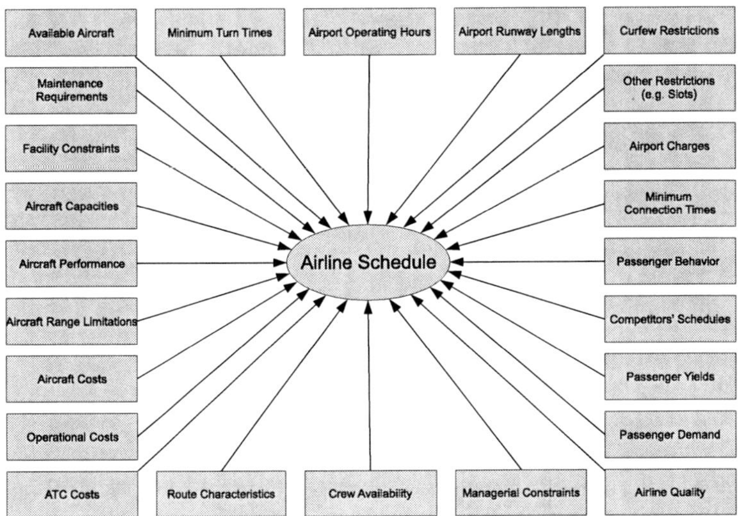

Fig. 2.1 Selection of input data for airline scheduling

2003). The subproblems are less complex and can be solved independently with (advanced) optimization approaches, each with its individual objective functions (Etschmaier & Mathaisel, 1985). In this process, the solution of one subproblem serves as input for the succeeding subproblem. Some of the subproblems are grouped together to build aggregate phases in the airline scheduling process. Numerous suggestions of researchers and practitioners exist on how to decompose the airline scheduling problem, how to aggregate subproblems, and how to order the subproblems within the scheduling process.[1] Although the decomposition of the overall problem and the order of the subproblems may vary among different solution approaches, each subproblem has to be solved to obtain an airline schedule that is feasible and can be operated (Suhl, 1995). In the following Fig. 2.2 only one example is given of a decomposition of the airline scheduling problem and an aggregation to three phases.[2]

For almost each of these subproblems, decision support systems (DSS) were developed. The process of airline scheduling then consists of an extensive and detailed interaction between human experts and DSS. The objective of these DSS is to simplify decision making through graphical user interfaces, to assess current schedules or elements, to check for feasibility and compliance with restrictions and regulations, and to process information through databases among the different planning

---

[1] For different approaches see for example Suhl (1995), Barnhart and Talluri (1997), Mathaisel (1997), Rushmeier and Kontogiorgis (1997), Grandeau et al. (1998), Gopalan and Talluri (1998b), Jarrah et al. (2000), Erdmann et al. (2001), Leibold (2001), Barnhart et al. (2003), Lohatepanont and Barnhart (2004).

[2] The proposed structure is inspired by various models and applications to the airline scheduling problem and its subproblems from scientific publications. A more detailed decomposition of the airline scheduling problem is presented by Antes (1998).

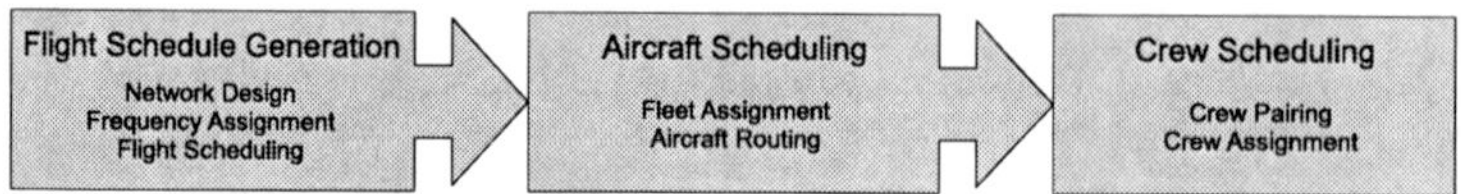

Fig. 2.2 Airline scheduling process

steps (Etschmaier & Mathaisel, 1985; Franken, 1990; Suhl, 1993; Suhl, 1995; Mathaisel, 1997; Rushmeier & Kontogiorgis, 1997; Grandeau et al., 1998; Kontogiorgis & Acharya, 1999). Furthermore, one of the most important tasks of DSS is to automatically solve subproblems of the airline scheduling problem. Many different optimization techniques from the field of operations research (OR) have found their way into DSS; in fact, airlines already began to identify the potential of OR methods to support their planning processes in the 1950s (Etschmaier & Mathaisel, 1985; Andersson, 1989) and have been active participants in OR since then. Since that time, a lot of different OR-models have been applied to various airline problems and especially the airline scheduling problem (Richter, 1989; Ball, 2003; Barnhart et al., 2003). Their implementation in DSS focuses on the subproblems that are well structured and, thus, can be formulated as mathematical optimization models.

Because of their success concerning selected subproblems which was due to advanced OR techniques and increased computational performance, much effort was and is undertaken to widen the scope of the methods of OR for airline scheduling. The level of detail an optimization model can represent and the number of decision variables solved in one step have constantly been increased. Furthermore, the field of application of OR methods was extended from simple and very well structured problems towards less structured problems. Thus, the extent to which automatic routines can be used in the scheduling process has constantly grown. As a result, the human interaction with DSS has shifted from operational decision making towards strategic planning and forecasting. However, since there is no single optimization model for the complete airline scheduling problem, there effort is still necessary to administer the complete airline scheduling process and to deal with interdependencies between single planning steps. Thus, although the individual models have been improved and extended, the airline scheduling process is still characterized by a sequential approach in which subproblems are solved step by step and complex and time-consuming (and, thus, expensive) feed-back loops and iterations are implemented to account for interdependencies and to improve the overall solution quality.

2.1.2 Outline

In the remainder of this chapter, an overview of the subproblems of the airline scheduling problem is given. Their objectives, inputs, and constraints are presented together with solution models from existing literature to solve them.[3]

---

[3] Similar (but less extensive) overviews can be found for example in Etschmaier and Mathaisel (1985), Teodorovic (1988), Suhl (1995), Barnhart and Talluri (1997), Rushmeier and Kontogiorgis (1997), Antes (1998), Gopalan and Talluri (1998b), Grandeau et al. (1998), Jarrah et al. (2000), Erdmann et al. (2001), Barnhart et al. (2003), Barth (2005).

After focusing on single subproblems, a presentation of solution approaches follows in Sect. 2.5 that integrate two or more subproblems to capture their interdependencies and, thus, to achieve a higher solution quality.[4]

The focus of this overview is on models used to create and optimize an airline schedule well ahead of the days of operation (tactical planning from the time-line perspective). Because the data and information used for this planning task might change until the day of operation, usually a continual refinement takes place. Thus, the closer the day of operation of a flight, the more modifications to the original schedule are necessary to improve the actual solution (Gopalan & Talluri, 1998b; Grandeau et al., 1998). Depending on the planning steps involved, this problem is referred to as re-scheduling or re-assignment.[5] A second need for modifications to a schedule is caused by irregular operations because of airport closures, weather effects, or unscheduled maintenance (Grandeau et al., 1998). In such a case, the problem is to keep the effect of these disturbances as low as possible and to return to regular operations as soon as possible after an exception.[6] Since the focus of this study is on the airline scheduling process described above, these problems are not considered in the remainder.

Furthermore, the focus is on contributions which are directly related to the passenger airline domain.[7] There are several publications regarding scheduling problems within air cargo or freight transportation, especially dealing with network design and flight scheduling problems. Although there are similarities between a passenger and a cargo airline, some major differences exist. For example, passengers are unwilling to make large detours on their journey caused by inconvenient connection flights. In contrast, detours are not a problem for cargo airlines as long as the shipment arrives on time (Chestler, 1985; O'Kelly, 1986). On the other hand, a cargo airline has to make shipment flow decisions, because deliveries between the same origin and destination can take different routes (Armacost et al., 2002; Link, 2006). Whereas for competing passenger airlines the departure time of each flight has a major impact on passenger demand, flight scheduling in cargo airlines only needs to ensure the offered service (for example overnight delivery). As cargo flights do not need a cabin crew, crew scheduling is less important in comparison to

---

[4] All presented models and citations are published and accessible. However, contributions to the airline scheduling problem were presented at annual conferences of the *Airline Group of the International Federation of OR Societies (AGIFORS)*. Since the proceedings of these meetings are inaccessible for the public, those contributions could not be considered in this study. For the same reason technical or internal company reports had to be excluded.

[5] For solution models see for example Teodorovic (1985), Yau (1989), Berge and Hopperstad (1993), Klincewicz and Rosenwein (1995), Teodorovic (1995), Talluri (1996), Stojkovic et al. (1998), Kontogiorgis and Acharya (1999), Jarrah et al. (2000), Moudani and Mora-Camino (2000), Stojkovic and Soumis (2001), Sriram and Haghani (2003).

[6] Some problems and models regarding the exception handling can be found in Argüello et al. (1997), Luo and Yu (1997), Wei et al. (1997), Yan and Tu (1997), Argüello et al. (1998), Clarke (1998), Luo and Yu (1998a), Luo and Yu (1998b), Song et al. (1998), (Lettovsky et al., 2000), Thengvall et al. (2000), Filar et al. (2001), Thengvall et al. (2001), Yu et al. (2003), Thengvall et al. (2003), Rosenberger et al. (2003), Sherali et al. (2006).

[7] For example, crew scheduling issues with similar properties arise in many different industries.

passenger airlines. Because of the differences between cargo and passenger airlines, the cargo domain is excluded in the remainder.[8]

Differences between a charter airline and an airline operating with a regular schedule mainly concern the consideration of passenger demand that has a major impact on airline schedule design decisions. In contrast to scheduled airlines, the market for charter airlines is well-known. Large contingents of seats are booked before the airline schedule generation process is terminated, which, in addition to the historical data on customer behavior, gives a very accurate knowledge about the market demand (Erdmann et al., 2001). Thus, the schedule of charter airlines usually does not represent its most important competitive marketing instrument, and factors like departure times or travel times play only a minor role. Charter airlines are considered in the papers of Desrosiers et al. (2000) and Erdmann et al. (2001).

Finally, some subproblems are related to or could be formulated as vehicle routing problems (VRP) or traveling salesman problems (TSP) and their extensions. A large number of approaches exist to solve these generic models that are out of the scope of this study.[9]

2.2 Flight Schedule Generation

2.2.1 Problem

The generation or construction of a flight schedule is the first phase in the airline scheduling process. It is one of the most important steps in airline schedule planning because it affects every subsequent planning step and has the largest impact on passenger demand (Barnhart & Talluri, 1997; Antes et al., 1998; Hsu & Wen, 2000; Erdmann et al., 2001). The objective of this planning step is to develop a schedule that is presented to the public and that includes exact information about the offered flights (Grandeau et al., 1998):

- departure and arrival airports (stopover airports in case of a connecting flight),
- departure and arrival times (determining travel times),
- flight frequencies, and
- days of operations.

The flight schedule generation is conducted between twelve weeks and six months prior to the actual operation of the schedule; the resulting schedule is usually fixed

---

[8] Surveys and optimization models in air cargo scheduling can be found for example in Marsten and Muller (1980), Chestler (1985), O'Kelly (1986), Crainic and Roy (1988), Hall (1989), Verwijmeren and Tilanus (1993), Kuby and Gray (1993), Barnhart and Schneur (1996), Kasilingam (1997a), Kasilingam (1997b), Mason (1997), Raguraman (1997), Antes et al. (1998), Kim et al. (1999), Morrell and Pilon (1999), Büdenbender et al. (2000), Crainic (2000), Armacost et al. (2002), Armacost et al. (2004), Link (2006).

[9] Surveys and problem-related overviews can be found for example in Bodin et al. (1983), Golden and Assad (1988), Solomon and Desrosiers (1988), Desrochers et al. (1990), Golden and Wong (1992), Desrosiers et al. (1995), Freling et al. (1997), Kim and Barnhart (1997), Desaulniers et al. (1998), Bard et al. (2002).

for a period of time, typically three or six months (Gopalan & Talluri, 1998b; Butchers et al., 2001; Taneja, 2002). However, in the construction phase airlines usually consider a pattern schedule for a shorter time interval (a day or a week), so that the cyclical extension of this schedule represents the flight schedule over the whole planning horizon (Feo & Bard, 1989; Grandeau et al., 1998; Mashford & Marksjo, 2001). This procedure has the advantage of complexity reduction in scheduling operational tasks and of increased simplicity for passengers.

The schedule is usually determined based on traffic forecasts, tactical and strategic initiatives, and seasonal demand variations (Gopalan & Talluri, 1998b; Taneja, 2002). Major constraints in this planning step are the size and composition of the aircraft fleet and other resources, and legal factors like traffic rights (Feo & Bard, 1989; Barnhart & Talluri, 1997; Grandeau et al., 1998; Antes et al., 1998). However, since the assignment of resources is conducted in later planning steps, these restrictions can only be considered roughly.

Within the phase of flight schedule generation, three subproblems can be identified:

1. network design,
2. frequency assignment, and
3. flight scheduling.

2.2.1.1 Network Design

The objective of the network design problem is to identify origin-destination city pairs (O&Ds or markets) that the airline intends to serve. Once the markets are selected, the airline has to decide about its route network. Each O&D can be carried out either by a direct flight (single-leg flight) or by a connection flight (multi-leg flight).

Two well known route network structures are the hub-and-spoke network and the point-to-point or direct-service network (see Fig. 2.3).

In a pure point-to-point network, nonstop flights exist between all cities, whereas in hub-and-spoke networks only larger (hub) cities are connected directly and cities with smaller demand (spokes) are connected only to a near hub. Passengers not traveling between hub cities have to change planes at hubs. In this network structure the airlines can consolidate passenger flows from several city-pair routes and combine these individual routes on hub flights, reducing the amount of variations in the number of passengers (Hsu & Wu, 1997; Lohatepanont & Barnhart, 2004). Thus, economies of flow concentration can be realized and the lowest possible operating costs can be achieved (Ghobrial & Kanafani, 1995a; Hsu & Wen, 2000). In contrast to a point-to-point network, more markets can be served with a higher frequency with the same number of aircraft (Lederer & Nambimadom, 1998; Patty & Diamond, 1998). The drawback of a hub-and-spoke system is the inconvenience for a traveler to have to use connection flights with increased total travel time and possible congestion at traffic peaks at the hub airport. However, the increased frequency can offset the passenger's longer travel times (Ghobrial & Kanafani, 1995a; Hsu & Wen, 2000).

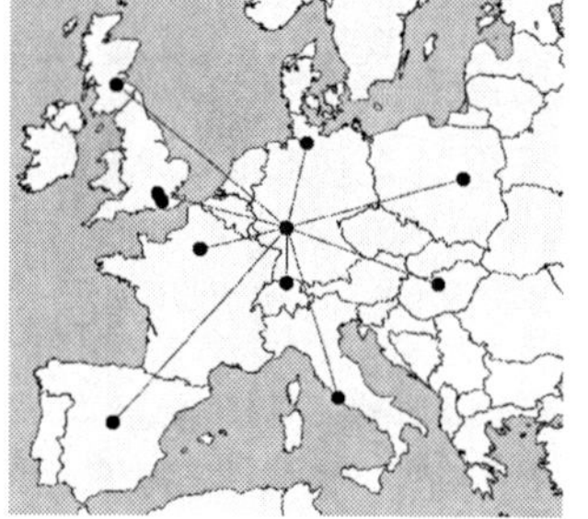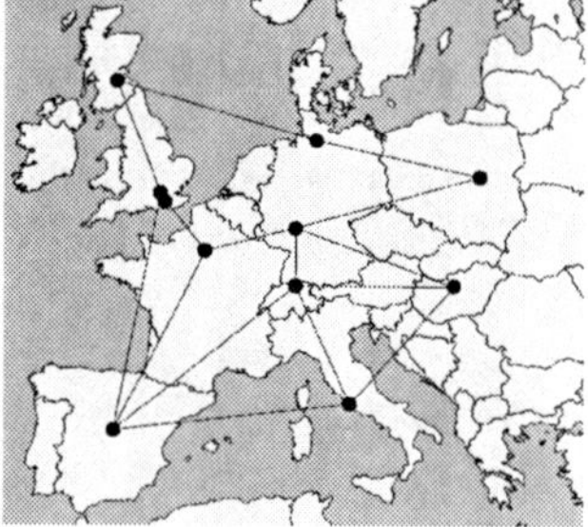

Fig. 2.3 Hub-and-spoke and direct-service network

The (two-stop) hub-and-spoke network became the most common network structure in the USA after the deregulation and typically consists of three to seven hubs (Jaillet et al., 1996; Lederer & Nambimadom, 1998; Reynolds-Feighan, 2001; Button, 2005). However, on average there are more direct connections today than before 1978 (Barnett et al., 1992). In addition, low-cost airlines usually have a point-to-point network (Alderighi et al., 2005), thus, both systems are established in coexistence. In fact, even for a single airline it is recognized that a mixed system is the most profitable network structure (Dobson & Lederer, 1993; Lederer & Nambimadom, 1998).

Airline networks are studied in several disciplines. Economics researchers, operations researchers, and transportation engineers have all made contributions to understand airline networks (Lederer & Nambimadom, 1998). Especially hub-and-spoke networks (and in some cases point-to-point networks as comparison) have attracted much attention.[10] These studies mainly include empirical *ex post* analyses of existing route networks and investigate simplified models with regard to their impact on service quality in terms of frequency, air fares etc. The external perspective in the studies reduces their applicability to the flight schedule generation problem. However, the major findings that can be used as a suggestion when choosing a route network might be summarized in one sentence: A point-to-point (hub-and-spoke) network is advisable if distances between cities are very large (small), demand is high (low), and there is a small (high) number of cities (Lederer & Nambimadom, 1998; Wojahn, 2002).

2.2.1.2 Frequency Assignment

The aim of this planning step is to assign a frequency (number of flights during the planning horizon) to each O&D. Some studies identified the frequency and the fare as the main driving forces for a passenger's choice of a flight or carrier. Because the air fare is in most cases determined by the intensity of competition on the

---

[10] See for example Oum and Tretheway (1990), Dennis (1994b), Dennis (1994a), Ghobrial and Kanafani (1995a), Jaillet et al. (1996), Hsu and Wu (1997), Bania et al. (1998), Lederer and Nambimadom (1998), Hsu and Wen (2000), Wei and Hansen (2006).

market, many researchers denote the frequency as the only possibility for an airline to differentiate from its competitors and to influence the demand for its flights.[11] The influence of frequency on passenger demand can be explained by two factors:

- An airline that offers a high number of flights in a market is perceived more and, thus, receives more attention from passengers and is more attractive for them. Some studies examined this relation between the frequency of an airline and its impacts and identified a disproportional relation between market share and frequency share (Gelerman & de Neufville, 1973; Bouamrene & Flavell, 1980).
- One important factor when traveling is the total travel time. The total travel time is the sum of the total flight time, time at a connection airport, access time at the departure and arrival airport, and the difference between the desired and the offered departure or arrival time. With an increasing frequency and, thus, number of departures, the difference between desired and offered departure time decreases, making each flight more attractive to passengers (Jorge-Calderón, 1997; Billette de Villemeur, 2004; Wei & Hansen, 2005).

2.2.1.3 Flight Scheduling

In the flight scheduling problem, exact departure times are determined. Some authors include the designation of the arrival times. However, because different aircraft types fly with different speeds and are assigned to the flights in a later planning stage, only approximate arrival times can be specified (Subramanian et al., 1994; Barnhart & Talluri, 1997). The choice of proper departure times depends mainly on two factors:

- *demand distribution*: The demand for air travel is not distributed uniformly over a given horizon (for example a week) but varies with the day of the week and time of the day. Because business travelers need to travel at the beginning and the end of regular business days and weeks, there is a higher demand at the beginning and end of each week and in the morning and evening of each day (Teodorovic & Krcmar-Nozic, 1989; Proussaloglou & Koppelman, 1999). The demand distribution and its peaks can be described by *time-of-week* or *time-of-day curves*.[12] However, passengers that travel for pleasure mainly focus on price when selecting a flight.
- *connection possibilities*: The main objective of the installation of hub-and-spoke networks is to offer a large number of O&D-itineraries. This can be accomplished by matching single flight legs to form connection flights. A minimum connection time has to be considered to allow a passenger to change aircraft. On the other hand, a passenger wants to reduce his total travel time, thus, stopovers should be as short as possible. These considerations have to be taken

---

[11] See for example Gelerman and de Neufville (1973), Bouamrene and Flavell (1980), Morrison and Winston (1985), Teodorovic and Krcmar-Nozic (1989), Ghobrial and Kanafani (1995a), Jorge-Calderón (1997), Borenstein and Netz (1999).

[12] See Fig. 4.12 on page 86 for examples of different time-of-day curves that were derived from observed travel behavior.

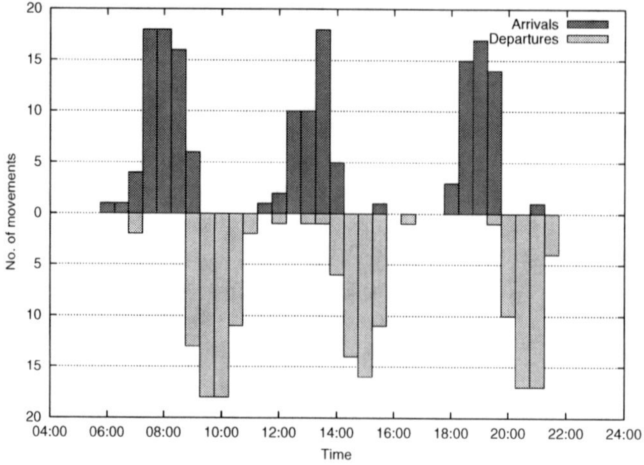

Fig. 2.4 Schedule structure of the Alitalia hub in Milan Malpensa (MXP) on 19th January 2005 (Source: Danesi (2006))

into account when selecting departure times for flight legs. The practice in hub-and-spoke networks is to schedule *waves*: in a short time interval, a high number of aircraft arrive or depart at the hub. Between two waves passengers can connect between flights. As a relatively large number of aircraft is on ground at the same time, many different connection flights can be offered (Dennis, 2000; Patty & Diamond, 1998). Fig. 2.4 gives an example of this concept, presenting the number of arriving and departing flights over the day at Milan Malpensa. Periods of arriving flights followed by many departures are clearly visible.

2.2.2 Solution Models

General network design and scheduling problems have been intensively studied in different disciplines. Many different algorithms and applications were developed. For surveys, see for example Magnanti and Wong (1984), Minoux (1989), Jaillet et al. (1996), Kim and Barnhart (1997), Gendron et al. (1999), Crainic (2000). However, since the flight schedule generation problem of airlines is far more complex, the typical airline practice today is to build flight schedules manually, with limited optimization. No airline uses an automated model that captures all the relevant factors and constructs a schedule from scratch (Barnhart & Talluri, 1997; Gopalan & Talluri, 1998b; Erdmann et al., 2001; Smith et al., 2001; Taneja, 2002; Yan & Tseng, 2002; Barnhart et al., 2003). With recent research advances, optimization is getting more attention also in the early planning stages (Barnhart et al., 2003). Simplified or heuristic techniques can be used to support a manual decision process when creating a schedule (Day & Ryan, 1997). In the following, a selection of models that can be used to support the flight schedule generation phase is presented.

2.2.2.1 Network Design

A lot of research on hub-and-spoke-networks has been conducted and models were developed that assist in making routing decisions within this network structure. A classical and frequently encountered problem is the single-hub location problem (with non-hub routes prohibited). The objective is to find a hub location so that each non-hub city is connected to the hub and total transportation costs are minimized. Campbell (1994b) provides a survey of network hub location problems. O'Kelly (1987) addresses this problem using a quadratic integer model with two heuristics. Integer programming formulations for a variety of single-hub and multi-hub location problems are presented by Campbell (1994a) and hub center and hub covering problems are introduced. In the model of Aykin (1995) direct connections are allowed. The author presents several integer programming models for single-hub and multi-hub location problems including fixed costs for establishing hubs. The proposed solution procedure includes enumeration algorithms and greedy-interchange heuristics. Jaillet et al. (1996) observe that candidates for hub locations depend more on their geographical position than on their own demand level.

Jeng (1988) studies the mix of hub and direct routing flights chosen by an airline, and shows how network parameters such as demand level, geographical distance between cities, and number of cities served affect the routing choices. Lederer (1993) and Marianov et al. (1999) consider competition when making routing decisions. In the model of Lederer (1993), competition is modeled as a non-cooperative game where airlines select network designs first and then prices for transportation between any two nodes. A game-theoretical approach assuming two carriers is also used by Alderighi et al. (2005); the authors identify conditions under which a hub-and-spoke and a point-to-point network can coexist. Marianov et al. (1999) use a heuristic approach. Their one- or two-hub model allows an airline to capture customers if it can provide a shorter distance (or time) from the origin to the destination.

2.2.2.2 Frequency Assignment

Hansen (1990) presents an n-player, non-cooperative game in which the airline's sole strategy set is frequency of service. The set of simplifying assumptions include fixed air fares, adequate capacity, inelastic demand of price and service level and consideration of nonstop and one-stop services only. Teodorovic and Krcmar-Nozic (1989) use a two-step approach to determine flight frequencies in an airline network under competitive conditions. The relation between frequency and the number of passengers is estimated by simulation in the first step. Then, the optimal frequency for each route is obtained by combining a heuristic algorithm with a multi-variable non-linear integer problem.

2.2.2.3 Flight Scheduling

Trietsch (1993) faces the problem of scheduling connections at a hub facility where arrivals and departures are subject to stochastic variation. Based on given optimal departure times, the scheduled ground time of each plane is adopted to minimize the expected sum of the airline's and passengers' costs and delay penalties. Wu and Caves (2002) work on a similar problem. Based on their earlier investigation (Wu & Caves, 2000) on how the trade-off situation between the ground time of a turnaround aircraft and schedule punctuality performance varies with the buffer time allocated to the schedule, it is their objective to improve the reliability and robustness of a flight schedule by implementing buffer time between two succeeding flights and, thus, scheduling the departure times. The same problem is tackled by Lan et al. (2006). Departure times are rescheduled within given time windows to reduce passenger disruptions because of flight delays without requiring additional aircraft to fly the schedule. Another approach to increase a schedule's robustness via rescheduling is presented by Lee et al. (2007). The problem is modeled as a multi-objective programming problem and solved using a genetic algorithm (GA). The use of a multi-objective genetic algorithm allows to choose different parameters as a measure for robustness. To evaluate the flight schedules, a simulation model is applied which simulates airline operations under operational irregularities.

In the approach of Patty and Diamond (1998), the flight scheduling problem for a special case is considered. In a single-hub network flights to and from spoke cities have to be assigned to waves at the hub airport. Each complex is specified by its directionality (inbound or outbound), a time interval, and the number of possible flights (which is restricted by the number of available gates or slots). The number of desired frequencies between two cities is given and the objective is to find a feasible assignment of flights to the complexes requiring the minimum number of aircraft while providing the desired level of service. This problem is formulated as a network model with side constraints and an enhanced primal partitioning approach is used to obtain optimal solutions.

2.3 Aircraft Scheduling

2.3.1 *Problem*

While OR has had relatively little impact on the flight schedule generation phase, the aircraft and crew scheduling phases have attracted more attention. A reason for this observation might be that these phases build on a given flight schedule whose construction already removed a large amount of uncertainty. Aircraft and crew scheduling problems are better structured and, thus, can better and more reliably be expressed in mathematical models.

The objective of the aircraft scheduling phase is to assign one aircraft to each flight leg of a given flight schedule. Two subproblems have to be solved in this phase: fleet assignment, and aircraft routing.

2.3.1.1 Fleet Assignment

The fleet assignment is usually solved several months before operating the flights, its solution affects all later planning steps in the airline scheduling process

(Barnhart & Talluri, 1997; Rexing et al., 2000). The objective of the fleet assignment problem is to assign aircraft types to the flight legs so that profit is maximized.

Airlines usually possess aircraft of several types. Aircraft of the same type (fleet) have some characteristics in common: cruising speed, fuel consumption, capacity, noise restrictions, crew requirements, required ground equipment, maintenance requirements, cost structures, and minimum turn times[13] (Barnhart et al., 1998; Gopalan & Talluri, 1998b; Wu & Caves, 2000). As a result, different fleets produce different revenues if assigned to the same flight segment, and, given a fleet schedule, a large part of the airline cost estimates and total revenue is fixed (Gu et al., 1994; Desaulniers et al., 1997).

In fleet assignment, profit is maximized by minimizing two types of costs: operational costs and spill costs (Subramanian et al., 1994; Barnhart et al., 1998; Barnhart et al., 2002). Operational costs are the costs for flying the flight leg with the assigned aircraft type and usually include such costs like fuel and landing fees. Spill costs represent opportunity costs that arise if passenger demand exceeds the aircraft capacity and, thus, potential revenue is lost (Barnhart et al., 2003).

A fleet assignment has to satisfy many constraints. Major constraints include:

- *aircraft count*: The number of assigned aircraft of one fleet may not exceed the number of available aircraft of this type.
- *flight coverage*: Each flight has to be covered by exactly one aircraft type.
- *flow balance*: The number of arrivals and the number of departures of each fleet at one airport have to be the same. If the flow balance constraint is not satisfied, flights without passengers are needed to reposition an aircraft (*deadhead* flights) and the problem becomes more complicated (Gopalan & Talluri, 1998b).
- *curfew restrictions*: A fleet's operational limitations at certain airports have to be considered (e.g. noise limitations, runway lengths, gate sizes etc.).

In the basic fleet assignment model, maintenance and crew constraints are not considered. However, because major interdependencies between these problems exist, a higher solution quality can be achieved by integrating maintenance and crew decisions in the fleet assignment problem. Such models are presented in Sect. 2.5.

For domestic service, the fleet assignment model is usually formulated for a typical day. On international routes or for a less regular schedule, the airline has to solve a more complicated weekly fleet assignment problem (Talluri, 1996; Barnhart & Talluri, 1997; Andersson et al., 1998; Gopalan & Talluri, 1998b; Emden-Weinert & Proksch, 1999; Jarrah et al., 2000). In a daily fleet assignment, modifications to the fleet schedule for weekend flights have to be made in a separate step in order to capture the different demand structures for these flights (Subramanian et al., 1994; Talluri, 1996; Gopalan & Talluri, 1998b; Kontogiorgis & Acharya, 1999; Jarrah et al., 2000).

Whereas a variable or weekly fleet assignment promises improvements in revenue, a daily fleet assignment makes operations easier (Barnhart & Talluri, 1997). It

---

[13] The minimum turn time denotes the time that is necessary to prepare an aircraft after landing for the next flight, operations during this time (usually 30-60 minutes) include (de-) boarding, cleaning, refueling etc.

is easier to schedule gates, crews and maintenance when the same equipment types are assigned to fly the same flight legs every day. Gopalan and Talluri (1998b) state that these benefits generally outweigh the additional revenue obtained by a variable fleeting. Although passenger demand varies by the day of week, the variation is usually system-wide. The demand is uniformly high over the entire system on certain days and uniformly low over the entire system on certain days. Therefore, it is not clear how much more demand can be captured on the high demand days.

2.3.1.2   Aircraft Routing

In the fleet assignment problem, only one aircraft type is assigned to each flight leg. Given this fleet schedule, the objective of the aircraft routing is to find a feasible and profit-maximizing assignment of physical aircraft to the flight legs (Barnhart et al., 1998). Because each individual aircraft can be identified by its registration or *tail number*, this planning step is sometimes referred to as tail routing (Barnhart & Talluri, 1997; Antes, 1998). The aircraft routing problem is usually conducted for each fleet separately (Barnhart et al., 1998).

An aircraft routing consists of a number of flight legs that can be operated by the same aircraft (thus, the departure time of a leg must be greater than the arrival and turn time of the previous leg). A sequence of aircraft routings that starts and ends at the same location and that can be flown by one aircraft is denoted as rotation. A rotation can be flown by more than one aircraft (in parallel), and many airlines construct one rotation per fleet (Gu et al., 1994; Clarke et al., 1997). Each aircraft then is subject to identical flying conditions and maintenance requirements, resulting in an equal utilization of the aircraft and easier operational planning.

In the aircraft routing problem two subproblems can be identified:

- through flight assignment, and
- maintenance routing.

Through Flight Assignment. Passengers are willing to pay a higher fare if they do not need to change aircraft on connecting flights because they do not need to transfer gates to make a connection and can avoid the possibility of irritations such as misrouted baggage (Gopalan & Talluri, 1998b).[14] This phenomenon is addressed in the through flight assignment problem: the objective is to identify flight legs that have the highest impact on profit when flown by the same (physical) aircraft. The higher revenue of such a sequence of flight legs (through flights) is expressed by *through values* associated with the routing.

Additionally, through flights are ranked above regular connection flights on the screens of computer reservations systems (CRS). Research has shown that the order

---

[14] The problem of assigning gates at airports to flights is related to this subject. In this problem, usually the total distance that every connecting passenger has to walk to change aircraft is minimized. Models and solution approaches for the gate assignment problem were developed by Mangoubi and Mathaisel (1985), Vanderstraeten and Bergeron (1988), Bihr (1990), Gosling (1990), Su and Srihari (1993), Cheng (1998), Haghani and Chen (1998), Yan and Chang (1998), Bolat (1999), Gu and Chung (1999), Bolat (2000), Yan and Huo (2001), Yan and Tang (2007).

Table 2.1 Maintenance system for a jet aircraft (hypothetical example) (Source: Wells (1998))

Inspection	Time between Inspections	Labor	Duration	Work Performed
En-route service	each stop	1 hour	0.5 hour	*walk-around* – visual inspection to ensure no obvious problems, such as leaks, missing rivets, or cracks
Overnight	8 hours	Varies	Up to 8 hours	Ad-hoc repairs – Work varies
A-check	125 hours	60 hours	8 hours	Primary examination – fuselage exterior, power plant, and accessible subsystems inspected
B-check	750 hours	200 hours	Overnight	Intermediate inspection – panels, cowlings, oil filters, and airframe examined
C-check	3,000 hours	2,000-12,000 hours	5 days	Detailed inspection – engines and components repaired, flight controls calibrated, and major internal mechanisms tested
D-check	20,000 hours	15,000-35,000 hours	15-30 days	Major reconditioning – cabin interiors removed, flight controls examined, fuel system probed, and more

of appearance has a high impact on the passengers' choice of a flight, thus, the airlines try to position their flight as far above as possible (Copeland et al., 1995; Suhl, 1995; Gopalan & Talluri, 1998b).[15]

Maintenance Routing. The aviation authorities of each country (for example the FAA in the USA) require each aircraft to undergo regular maintenance checks. The requirement of a maintenance check usually depends on a combination of flight hours and landing and take-off cycles of each aircraft (Sriram & Haghani, 2003). In general, four categories of maintenance checks can be identified: the A-, B-, C-, and D-check. For example, the A-check has to be conducted every 65 flight hours or about once a week and has a regular duration of about eight hours, whereas the heavier C- or D-check occurs every one to four years taking the aircraft out of service for up to a month (Sriram & Haghani, 2003). Table 2.1 presents an (hypothetical) example for the required maintenance checks of a jet aircraft.

Because the latter checks generally reduce the total number of aircraft available for schedule operations, in airline scheduling the main concern is to meet the A-checks (Feo & Bard, 1989; Clarke et al., 1997; Gopalan & Talluri, 1998b). For that reason airlines usually require each aircraft to undergo maintenance every three or four days, depending on the aircraft age (Talluri, 1998). Unless exceptional circumstances occur, inspections and repairs take place at night, ensuring a high aircraft

---

[15] For the problem of assigning flight numbers to the legs in the current schedule see for example the contribution of Gopalan and Talluri (1998b).

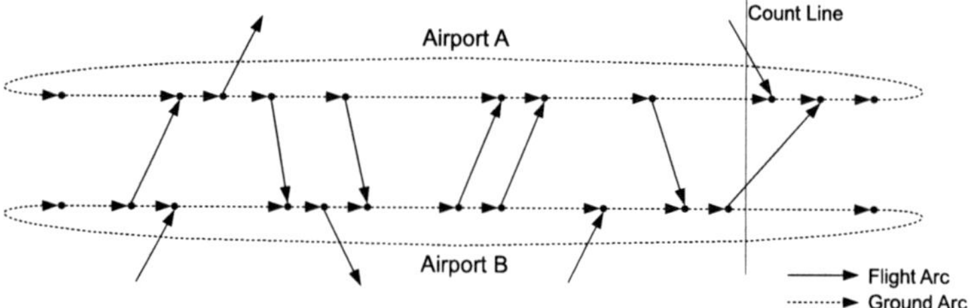

Fig. 2.5 Timeline network involving two airports (Source: Barnhart et al. (2003))

utilization (Feo & Bard, 1989; Subramanian et al., 1994; Barnhart & Talluri, 1997; Gopalan & Talluri, 1998a; Sriram & Haghani, 2003).

2.3.2 Solution Models

Aircraft routing and scheduling models were the earliest OR-models of airline planning. First approaches were developed in the 1950s (Ferguson & Dantzig, 1956b; Ferguson & Dantzig, 1956a) and 1960s (Dantzig, 1963; Miller, 1967). Most of these models consider simplified flight schedules where the number of alternative routes is small. The minimization of the fleet size necessary to fly the schedule was one common objective in those approaches (Pollack, 1974; Gertsbach & Gurevich, 1977).

2.3.2.1 Fleet Assignment

The complexity of the basic daily fleet assignment problem is studied by Gu et al. (1994). The authors analyze the structure of the solution as a function of the number of fleets; they observe that the complexity of the feasibility problem for two fleets is unknown and for three fleets it is *NP*-complete.

Most models to solve the daily fleet assignment problem are formulated as large multi-commodity flow problems with side constraints defined on a time-expanded network (Abara, 1989; Hane et al., 1995; Clarke et al., 1996; Barnhart & Talluri, 1997; Kim & Barnhart, 1997). The network contains flight arcs corresponding to flight legs, ground arcs corresponding to aircraft on the ground and overnight arcs corresponding to overnighting aircraft, the aircraft correspond to the commodities, a count line is used to determine the number of aircraft in use (see Fig. 2.5).

A typical mathematical formulation is given in the paper by Hane et al. (1995). However, as the number of integer variables is large, it can be difficult and time-consuming to find optimal integer solutions. These problems are often severely degenerated, which leads to poor performances of standard integer linear programming (LP) techniques. Hane et al. (1995) discuss various methods to decrease the size of the problem, including variable aggregation, cost perturbations, dual simplex with steepest-edge pricing, and intelligent branch-and-bound strategies. Most approaches to the fleet assignment problem focus on different techniques to reduce complexity when solving the formulated model (Sriram & Haghani, 2003).

For example, in one of the first models, Abara (1989) uses the simplex method, fractional variables are then rounded to obtain an integer solution.

Daskin and Panayotopoulos (1989) present an integer program to solve the fleet assignment problem for a hub-and-spoke network with a single hub. They propose a Lagrangian relaxation of the problem and combine it with heuristics for converting the Lagrangian solutions into primal feasible solutions.

Berge and Hopperstad (1993) address the re-fleeting problem. The mathematical programming formulation they describe is similar to the daily fleet assignment formulation, except for the aircraft count constraint, which is no longer required because the aircraft are positioned at the beginning of the planning horizon. The number of aircraft of each type present at each station at the beginning of the planning horizon ist fixed. They present two heuristic solution approaches: one heuristic solves a sequence of single-commodity flow problems, and the other begins with a feasible assignment and performs multiple profit-improving aircraft swaps.

A weekly fleet assignment model is solved by Kliewer and Tschöke (2000). The authors use a simulated annealing (SA) approach to deal with the higher complexity. Kliewer (2000) combines this approach with a demand model: Once one temperature level is completed, the current fleeting is sent to the market model. The market model then forecasts the number of passengers on each itinerary based on the given fleet assignment, and the simulated annealing algorithm continues.

In most fleet assignment models spill costs are leg-based. Thus, it is assumed that capacity is constrained only on the leg for which the estimate is being made and unconstrained on every other flight leg. In consequence, estimates of recaptured revenue are achieved without knowledge of capacity or passenger flow on the flight network. Barnhart et al. (2002) develop a fleet assignment model based on O&D-itineraries using a a branch-and-price approach. This model is capable of capturing network effects and more accurately estimating spill and recapture of passengers. Moreover, the authors include demand and fares for different fare classes in their model. Theoretically branch-and-price offers the best chance of finding a solution that is close to the optimum, but column generation requires the solution of a constrained shortest path problem which can be both memory and time consuming. Moreover, its application requires significant customization of the IP solver and best reduced cost columns may improve the LP value but not the IP value (Klabjan et al., 2001b). Relaxations, heuristic procedures or integration of domain-knowledge are common practices to support or replace these decisions (Anbil et al., 1992; Chu et al., 1997).

Further information on the fleet assignment problem including models and approaches is published by Sherali et al. (2006). In this paper, the authors present a tutorial including basic and integrated fleet assignment models on a detailed level.

2.3.2.2 Aircraft Routing

A common formulation for the aircraft routing problem is a network circulation problem with side constraints where exact and heuristic algorithms are applied to find feasible subtours. Solutions are considered to be feasible if each aircraft overnights at an appropriately equipped maintenance station at least every three or four days (Barnhart & Talluri, 1997). Thus, flight connections during the day are fixed and only overnights are allowed as maintenance opportunities (Gopalan & Talluri, 1998a). Because this may lead to aircraft rotations that are not able to fulfill the three- or four-day maintenance requirement, swapping techniques for the flights are necessary to unlock a rotation. Talluri (1998) develops a model for the four-day aircraft maintenance routing problem. Several heuristics and one exact approach are proposed to solve this problem. Furthermore, the mathematical complexity regarding the four-day routing problem and the three-day routing problem is investigated.

Bard and Cunningham (1987) consider the single through flight assignment problem. When through values are not considered, the aircraft routing problem is usually reduced to a feasibility problem (Cordeau et al., 2001; Klabjan et al., 2002). However, if through flight assignment and maintenance routing are solved separately, the latter problem is constrained by the results of the first, making it more difficult to find an optimal or even feasible solution (Gopalan & Talluri, 1998b). For example, since the pair of flight legs that constitute the through flight must be flown by the same aircraft, less freedom at the routing phase for the design of efficient routings that meet maintenance requirements is provided. Clarke et al. (1997) present an aircraft routing problem under consideration of through revenues and maintenance constraints. The objective is to build rotations that are profitable measured by the sum of through values of routing flights through airports, operationally attractive in terms of a single rotation for each fleet, and satisfy maintenance requirements by allowing aircraft to visit maintenance stations regularly for a sufficient length of time. In this approach, all connections between flights are used as options for maintenance (instead of using only overnight connections). Moreover, this approach can handle different types of maintenance requirements. The problem is formulated as an asymmetric traveling salesman problem with side constraints and is solved by using Lagrangian relaxation and heuristics.

Generally, in aircraft routing models capacity constraints at maintenance stations are not considered because A-checks usually only require minor inspections.[16] Some aircraft maintenance models consider the more intensive but less frequent *balance-check*. In order to ease scheduling operations, this constraint is met by performing the balance-check every n days whether there is one rotation of n aircraft in the fleet (Barnhart & Talluri, 1997; Gopalan & Talluri, 1998a).

Feo and Bard (1989) consider the maintenance location problem which involves finding the minimum number of maintenance stations required to meet the maintenance requirement for a proposed flight schedule. Stops during the day are not considered as maintenance opportunities. This problem is formulated as a minimum cost, multi-commodity network flow problem with integer restrictions on the variables. Because the size of the formulation is too large to optimize, they give a two-phase heuristic that begins by generating aircraft assignments to match flight requirements. A probabilistically set covering heuristic is then used to locate the maintenance stations.

---

[16] Duffuaa and Andijani (1999) present an integrated simulation model for the planning of maintenance operations for a maintenance station.

Lan et al. (2006) solve the aircraft maintenance routing problem to create robust schedules. The objective is to find an aircraft routing that is less susceptible to flight delays and does not propagate a single disruption through the following flights. The problem is formulated as a mixed-integer programming problem with stochastically generated inputs, the objective function is to minimize the expected total propagated delay for the selected routings.

Sarac et al. (2006) solve an operational aircraft maintenance routing problem. In this problem, a routing for individual aircraft to maintenance stations has to be found on a daily basis. In contrast to traditional maintenance routing approaches, this approach does not construct a regular maintenance schedule but focuses on the operational execution of maintenance, which might be affected by stochastic events. In addition, maintenance resource availability constraints are taken into account. Based on the remaining legal flying hours without violating maintenance restrictions, the aircraft have to be rerouted to appropriate maintenance stations with enough maintenance hours and maintenance slots. The problem is formulated as a set-partitioning problem and solved using a branch-and-price approach.

2.4 Crew Scheduling

2.4.1 Problem

Crew costs represent the highest direct operating cost of an airline after aircraft-related costs. Thus, significant cost reductions can be reached by solving the crew scheduling problem optimally (Anbil et al., 1991; Graves et al., 1993; Desaulniers et al., 1997; Andersson et al., 1998; Emden-Weinert & Proksch, 1999; Lucic & Teodorovic, 1999; Barnhart et al., 2003). Although a large number of airline personnel is necessary to operate an airline schedule, in traditional crew scheduling only aircraft crews are considered. An aircraft crew consists of the flight deck crew (pilot, co-pilot, and flight engineer on older planes) and the cabin crew (purser and flight attendants).

The objective of the crew scheduling problem is to find an optimal assignment of crew members to the flights determined by the previous planning steps. The quality of a solution is usually defined by a combination of crew costs and crew satisfaction (Lucic & Teodorovic, 1999; Gamache et al., 1999; Butchers et al., 2001; Sriram & Haghani, 2003). The crew scheduling represents a very complex optimization problem because of many constraints defined by work-rules that are given by legal regulations, union agreements, and company policies (Graves et al., 1993). Usually, a crew schedule has a duration of one month and is constructed four weeks to two months before operation (Anbil et al., 1991; Dillon & Kontogiorgis, 1999; Butchers et al., 2001; Cordeau et al., 2001).

Because of the complex structure of work-rules and crew costs, the crew scheduling problem is solved in a two-step process (Gamache & Soumis, 1998; Emden-Weinert & Proksch, 1999; Dawid et al., 2001; Barnhart et al., 2003; Guo et al., 2006):

1. crew pairing, and
2. crew assignment.

2.4.1.1 Crew Pairing

Given a flight schedule and fleet assignment, sequences of flight legs that can be flown by the same crew are generated (Vance et al., 1997; Cordeau et al., 2001; Yan et al., 2002). A sequence of flight legs with short rest periods separating them and brief and debrief times is called a duty period. The length of a duty period is usually one day. A sequence of duty periods with overnight rests between them forms a crew pairing (see Fig. 2.6). Each pairing begins and ends at the same crew base (cities where crews are stationed) and typically lasts between two and five days for domestic flights and up to three weeks for long-haul flights (Anbil et al., 1991; Hoffman & Padberg, 1993; Jarrah & Diamond, 1997; Desaulniers et al., 1997; Emden-Weinert & Proksch, 1999). However, in some cases a pairing includes flights that the crew fly as passengers (*deadhead* flights). These flights are used to reposition a crew to a city where it is needed to cover a flight, or to enable the crew to return to its base at the end of a pairing (Vance et al., 1997). The situation in which a crew has to stay outside its home base is referred to as *lonely overnight*. Although the costs of deadhead-flights are very high, there are cases where this can be economically feasible (Arabeyre et al., 1969). Especially in long-haul operations where relatively few flights may be scheduled in and out of a particular location, deadheading is an essential component (Barnhart & Shenoi, 1998). When allowing deadheading, the crew pairing problem becomes more complicated (Gopalan & Talluri, 1998b).

The objective of the crew pairing problem is to find a set of feasible pairings that minimizes crew costs and maximizes crew utilization (Jarrah & Diamond, 1997; Day & Ryan, 1997; Desaulniers et al., 1997; Butchers et al., 2001; Yan et al., 2002). In general, the planned pairings have to be feasible but are not allocated to individual crew members at this first stage (Day & Ryan, 1997; Butchers et al., 2001; Yan et al., 2002). One major constraint is the permission for each crew member to work only on certain aircraft types. A member of the flight deck crew usually is authorized

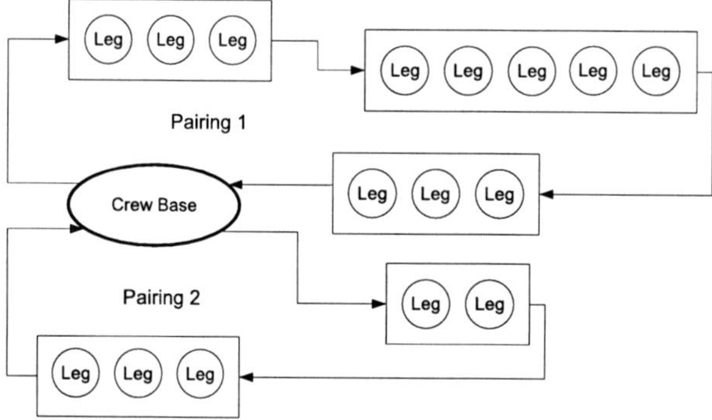

Fig. 2.6 Pairings in crew scheduling (Source: Suhl (1995))

to fly only one aircraft type, whereas cabin crew members usually work on different fleet types. In addition, the number of required pilots depends on the aircraft type whereas the number of flight attendants is specified by the number of expected passengers on a flight. Thus, the crew pairing problem is solved separately for the flight deck and cabin crew (Barnhart & Talluri, 1997; Vance et al., 1997; Andersson et al., 1998; Lucic & Teodorovic, 1999; Butchers et al., 2001; Sriram & Haghani, 2003). However, since the optimization problem remains the same, the two crew types are not distinguished in the remainder. Further restrictions in the crew pairing problem include (Arabeyre et al., 1969; Vance et al., 1997; Anbil et al., 1998; Lucic & Teodorovic, 1999; Butchers et al., 2001; Taneja, 2002; Barnhart et al., 2003):

- limits on the maximum number of hours worked in a day,
- limits on the maximum flying hours,
- limits on the maximum number of hours worked over several days,
- the minimum and maximum number of hours of rest between duty periods,
- the maximum number of flight duties in one pairing,
- the minimum and maximum number of hours of rest between flight legs,
- the maximum time the crew may be away from their home base.

Sometimes contractual obligations also require that the total flying amount has to be divided among the flight crews at different crew bases so that the total amount of flying hours assigned to crews from a given base must be within a specific interval. These restrictions ensure that crews at the various bases will all have the opportunity to receive credit for approximately the same number of hours of work each month (Vance et al., 1997). A further important feature of crew scheduling solutions from the management point of view is their operational robustness or sensitivity to disruptions of the planned flight schedule (Butchers et al., 2001; Yen & Birge, 2006). Often, operational robustness rules are sensible rules of thumb that minimize the impacts of disruptions on the day of operation. For example, providing rest periods slightly longer that the minimum legally required could allow crews to still have their legal rest periods if their flight arrives late. In addition, a desirable feature of a pairing may be that all crew members on a flight perform the same sequence of flights for as much of their duty period as possible and stay with the same aircraft. This greatly reduces the propagation of disruptions from one flight to other flights on the day of operation (Anbil et al., 1991; Butchers et al., 2001).

Like the restrictions, the costs of a pairing depend on agreements within the airline and, thus, may differ from airline to airline (Yan et al., 2002). A general cost structure may include the following variables (Barnhart & Talluri, 1997; Vance et al., 1997; Cordeau et al., 2001; Barnhart et al., 2003):

- costs per flying hour,
- accommodation expenses such as transport, meals, and hotel rooms when overnight connections take place outside the crew base,
- total elapsed work time,
- total time away from base,
- on-duty-time costs,

- deadhead costs,
- minimum guarantee costs (per day, per flight etc.).

Airlines with a stable and regular flight schedule solve a standard daily or weekly problem. Airlines with variable flight schedules must solve a fully dated problem, and the crew pairings solutions can differ from day to day (Andersson et al., 1998; Butchers et al., 2001).

2.4.1.2 Crew Assignment

In the previous step, only generic crew pairings are generated (Day & Ryan, 1997; Butchers et al., 2001; Yan et al., 2002). Individual crew members are assigned to the specified pairings in the crew assignment phase that is usually completed two weeks before operation (Butchers et al., 2001). After this step, each crew member has a personal flight schedule with the duties for the next planning horizon. Each work schedule usually consists of four to five pairings and typically lasts up to a month (Jones, 1989; Anbil et al., 1991; Dawid et al., 2001). As the pairings from the previous step serve as input for planning this problem, the crew assignment can usually be solved for each aircraft type and home base separately (Desaulniers et al., 1997; Emden-Weinert & Proksch, 1999; Sriram & Haghani, 2003).

The objective of the crew assignment problem is to find efficient and equitable assignments (Day & Ryan, 1997; Desaulniers et al., 1997; Gamache & Soumis, 1998; Gamache et al., 1999; Barnhart et al., 2003). This usually involves a conflict of management and crew objectives (Yan et al., 2002): management wishes to minimize the number of crew members required, while crew members wish to maximize their satisfaction with the roster. Although the crew scheduling problem has an obvious cost significance, the important issue of crew satisfaction also affects the costs of operating crew schedules (Day & Ryan, 1997). Crew dissatisfaction can indirectly lead to significant cost increases.

In crew assignment, two procedures can be identified:

- bidline generation, and
- rostering.

Bidline Generation. Bidline generation is the common practice of most US airlines. In this method, sequences of pairings that can be flown by one crew member construct bidlines (Jarrah & Diamond, 1997; Dillon & Kontogiorgis, 1999). Each bidline has to satisfy restrictions like minimum rest periods between each pairing, etc. The objective of the bidline generation step is to maximize the utilization of each crew member and some quality measure defined by the particular crew group (Christou et al., 1999; Butchers et al., 2001).

After the bidline generation process, individual crew members bid for each monthly work schedule according to their needs and desires. The bidding process is based on seniority, thus, a crew member with a higher rank and longer time of employment at the airline is more likely to be assigned to a favorable schedule (Jarrah & Diamond, 1997; Dillon & Kontogiorgis, 1999; Dawid et al., 2001).

Sometimes bidlines or pairings remain unassigned (Dillon & Kontogiorgis, 1999; Butchers et al., 2001). These flights, along with charters and unplanned flights are carried out by reserve flight crews (Dillon & Kontogiorgis, 1999). Reserve crew members do not select their flying in advance. Instead they bid for bidlines that consist of days on call.[17]

Rostering. As in bidline generation, pairings are grouped together to form monthly work schedules (rosters) for individual crew members. In contrast to bidline generation, rosters are assigned directly to the crew members. This problem is more complex than the bidline generation because apart from legal restrictions like minimum rest times etc. the availability or pre-assigned activities of each crew member have to be taken into account (Gamache & Soumis, 1998; Emden-Weinert & Proksch, 1999; Gamache et al., 1999; Lucic & Teodorovic, 1999; Butchers et al., 2001; Dawid et al., 2001; Yan et al., 2002):

- annual leave,
- training or observer flights,
- sick leave,
- visa regulations,
- transition activities that began in the preceding month and end at the beginning of the current month,
- medical appointments,
- flights or rest periods granted to an employee at specific times during the month.

In a *preferential bidding* approach, personalized schedules are constructed as in regular rostering, while also considering a set of crew members' bids for single flights (Gamache et al., 1998; Butchers et al., 2001). The aim is to maximize the award of crew members' preferences while respecting seniority.

The objective when building crew rosters is a fair and even distribution of the work load and unfavorable flights among all crew members and the maximization of the crew members' aggregated satisfaction, measured by various agreed collective quality measures (Gamache et al., 1999; Dawid et al., 2001; Butchers et al., 2001).

2.4.2 Solution Models

The crew scheduling problem is one of the most studied problems within the airline scheduling process. Moreover, many models applied to crew scheduling problems outside the airline domain can be transferred into airline scheduling. However, since work-rules and crew cost structures differ from airline to airline and from country to country, most solution approach models generalize crew scheduling problems. Andersson et al. (1998) present an overview of some similarities and differences in crew scheduling between Europe and North America.[18] An overview of commercial

---

[17] Dillon and Kontogiorgis (1999) and Sohoni et al. (2004) present an approach for the reserve crew scheduling problem.

[18] Surveys can be found in Arabeyre et al. (1969), Bodin et al. (1983), Ball and Roberts (1985), Gershkoff (1989), Barutt and Hull (1990), Jarrah et al. (1994), Desaulniers et al. (1998), Butchers et al. (2001), Barnhart et al. (2003).

systems used by major airlines for crew scheduling is given by Graves et al. (1993) and Desaulniers et al. (1997).

First solution approaches for airline crew scheduling problems were developed in the 1950s and 1960s (Arabeyre et al., 1969; Barnhart et al., 2003). Initially, heuristics were used because of the high complexity of crew scheduling problems and the lack of computer power; exact algorithms could only be applied to restricted problems (Arabeyre et al., 1969; Butchers et al., 2001; Barnhart et al., 2003). Since the mid-80s, improvements in optimization techniques and increasing computer power have resulted in the development of exact optimization methods for the crew scheduling problem (Butchers et al., 2001; Barnhart et al., 2003). Since then, researches have attempted both heuristic and exact approaches to solve the crew scheduling problem (Jarrah & Diamond, 1997; Barnhart et al., 2003). However, the crew scheduling process remains a very complex problem and even now many airlines still use either heuristic or manual methods to solve crew scheduling problems (Kwok et al., 1995; Butchers et al., 2001).

Both subproblems of crew scheduling, crew pairing and crew assignment, are usually formulated as set partitioning problems (SPP), set covering problems (SCP) or network models (Desaulniers et al., 1997; Andersson et al., 1998; Dillon & Kontogiorgis, 1999; Ozdemir & Mohan, 2001; Yan & Tu, 2002; Barnhart et al., 2003). As the number of combinations of trips is extremely large, researchers usually employ a branch-and-bound search tree in each problem (Dillon & Kontogiorgis, 1999; Barnhart et al., 2003). Most of the research has been focused on the improvement of these models and algorithms regarding branching and bounding decisions (Klabjan et al., 2001b; Yan & Tu, 2002; Yan et al., 2002). Those enhancements include for example constraint relaxation, heuristic procedures, decomposition approaches, integration of domain-knowledge, and column generation etc. (Barnhart et al., 1998; Barnhart et al., 2003). Although these approaches could efficiently optimize crew scheduling problems, when the problem size is increased or additional constraints are included, traditional SPP or SCP become more complicated and more difficult to solve (Yan & Tu, 2002)

2.4.2.1 Crew Pairing

In contrast to crew assignment, the crew pairing problem is computationally more intensive. In addition, crew costs are affected much more by the quality of the pairing than by the assignment (Anbil et al., 1991). Thus, the crew pairing problem has received more attention from researches than the crew assignment problem (Butchers et al., 2001; Dawid et al., 2001).

As stated, early attempts to solve this problem were based on heuristic procedures. For example, Baker et al. (1979) and Ball and Roberts (1985) use (simple) exchange procedures to iteratively improve a crew pairing solution. Barnhart et al. (1995) use a heuristic to improve crew pairing solutions through the efficient selection and utilization of deadhead flights.

In general, the crew pairing problem consists of two major components (Hoffman & Padberg, 1993): the generation as well as the optimization of feasible pairings.

Gershkoff (1989), Anbil et al. (1991), and Graves et al. (1993) propose a heuristic approach for the crew pairing problem consisting of these two steps: pairings are generated and then followed by an SPP for randomly selecting manageable subsets of the flight schedule by using an integer programming approach. Both steps are iteratively repeated until there are no improvements in the cost function.

Most approaches to crew pairing optimization use an SPP as problem formulation, in which the rows of a binary matrix correspond to the flight legs that have to be assigned to pairings. The columns correspond to feasible pairings. The objective is to find a good subset of pairings that cover each flight leg exactly once. The SPP is NP-hard and, thus, no efficient (polynomial) solution algorithm is available. Even for small instances, the number of pairings cannot be enumerated in reasonable time and for realistic problem sizes the set of possible pairings is innumerable (Anbil et al., 1991; Andersson et al., 1998). For example, in an early attempt using an SPP, the approach of Marsten and Shepardson (1981) is restricted to relatively small problems where the set of feasible pairings can be generated a priori. Much research has been conducted to overcome these difficulties. The majority of publications on the crew pairing problem mainly differ in the method of establishing a smart enumeration technique (Andersson et al., 1998). For example, the column generation approach proposed by Lavoie et al. (1988) is used in different variants in many publications.[19] A common practice is to combine randomness and domain knowledge when generating columns (Anbil et al., 1992; Chu et al., 1997). Barnhart and Shenoi (1998) suggest an approach that solves first an approximate model of the crew pairing problem. This near feasible advanced start solution then provides good lower bounds that speed up a column generation technique. This approach is designed for long-haul flight networks because they are relatively sparse and typically consist of a weekly schedule. Vance et al. (1997) solve the daily crew pairing problem in a two step process. In a first step good duty periods are selected that cover the flight schedule. A duty period represents a sequence of flight legs with short rest periods. These duty periods are used as building blocks in the second step where pairings are generated using a dynamic column generation approach. Klabjan et al. (2001b) first solve the linear programming relaxation and then select columns which best reduce the cost for the integer program. The number of columns is reduced by a heuristic based on linear programming. Finally an integer solution is obtained with a commercial integer programming solver. The branching rule of the solver is enhanced by combining strong branching and a specialized branching rule. Desaulniers et al. (1997) present a column generation approach for a weekly pairing problem that explicitly manages all rules and restrictions, including those related to duties, during the construction of feasible pairings.

An alternative approach without column generation is proposed by Hoffman and Padberg (1993). The authors present an exact branch-and-cut approach that is based on a heuristic to obtain good integer-feasible solutions quickly and a cut generation procedure to tighten the linear relaxation. In their approach, crew base constraints were explicitly considered. Mingozzi et al. (1999) use a heuristic procedure to find

---

[19] See for example Crainic and Rousseau (1987), Barnhart et al. (1994), Barnhart et al. (1998), Andersson et al. (1998), Anbil et al. (1998), Yan and Chang (2002).

lower bounds within a branch-and-bound approach to the SPP. Levine (1996) solves the SPP using a steady-state genetic algorithm (GA) that is combined with a local search heuristic. However, a traditional (specialized) branch-and-cut solution approach for the investigated problem instances was significantly more successful than the GA. A GA-approach is also used by Ozdemir and Mohan (2001). Instead of solving an SPP, the authors apply the algorithm to a flight graph representation that represents several problem-specific constraints. As input they use the flight schedule rather than pre-processed columns.

Desaulniers et al. (1997) formulate the crew pairing problem as an integer, non-linear multi-commodity network flow problem with additional resource variables. A branch-and-bound algorithm based on an extension of the Dantzig-Wolfe decomposition principle is used to solve this model. A network representation is also used by Yan and Tu (2002).

In the approach of Lagerholm et al. (2000) a feedback artificial neural network (ANN) is presented to solve the crew pairing problem. The objective is to minimize the total crew waiting time. One constraint limits the maximum number of flight legs in a pairing. A second constraint allows the crews only to follow rotations that start and end at the home base.

Usually, a daily crew pairing problem is solved. Klabjan et al. (2001a) solve the crew pairing problem for a weekly schedule. The authors extend the objective function of minimizing costs to capture the repetition or regularity of crew itineraries. Regularity is important with respect to crew (and aircraft) schedules, since regular solutions are much easier to implement and manage, and, if possible, crews prefer to repeat itineraries. The authors present a solution algorithm that integrates a heuristic in the consideration of subsets of columns with an optimization-based pricing approach.

In general, crew pairing for domestic or regional flight operations is more complicated than for international flights because domestic flights are shorter, and, thus, more flights are conducted within a given time interval. Hence, the number of possible connections a crew can make and the number of possible pairings is much larger than in international operations, especially when operating a hub-and-spoke network (Vance et al., 1997; Desaulniers et al., 1997; Andersson et al., 1998; Chang, 2002).

2.4.2.2 Crew Assignment

In contrast to the crew pairing problem, less work has been reported on solution methods for crew assignment (Butchers et al., 2001; Dawid et al., 2001). This step is still performed manually or with simple heuristics in many airlines (Dillon & Kontogiorgis, 1999; Gamache et al., 1999; Lucic & Teodorovic, 1999; Dawid et al., 2001; Butchers et al., 2001). One reason for this might be the difficulty of integrating crew satisfaction in a mathematical optimization model.

Bidline Generation. Jones (1989) presents a system for the monthly bidline generation problem. Using a heuristic technique, bidlines are generated that are deemed attractive by the crew members. To minimize the chances of missed flights due to crew member error, the numbers of days on and off are evenly distributed. Christou et al. (1999) address the bidline generation problem by using a genetic algorithm together with a local improvement heuristic that swaps assignments. Campbell et al. (1997) propose a simulated annealing approach. The objective is to minimize the number of bidlines and of unassigned flights (that would have to be assigned in subsequent steps). In a two-step process, first as many valid bidlines as possible are generated using simulated annealing, and second a greedy heuristic forces as many unassigned flights as possible onto additional valid bidlines.

Rostering. A simple heuristic for the crew rostering problem is presented by Nicoletti (1975). For each day of the month, pairings are assigned to individuals selected from a pool of available crew members. Thus, rosters are constructed day-by-day. In this approach, potential difficulties on succeeding days are not taken into account. Some pairings may not be assigned or uneven workloads are produced (Lucic & Teodorovic, 1999).

Lucic and Teodorovic (1999) present a simulated annealing approach for the rostering problem. Their approach attempts to solve the aircrew rostering problem as a multi-objective optimization problem, thus constructing personalized monthly schedules on the basis of several criteria.

Ryan (1992) and Gamache et al. (1999) formulate a basic model for the crew rostering problem as generalized SPP with side constraints. In the approach of Ryan (1992), a heuristic is first used to generate a priori a set of feasible rosters for each employee. The rest of the problem is then solved by using specialized integer programming. The objective of this approach is to minimize the number of crew members and to maximize a function that attempts to measure crew preferences. In the approach of Gamache et al. (1999) a heuristic is used to find a good integer solution. At each node of the decision tree, a column generation technique is used to solve the linear relaxation of the generalized SPP. The objective is to maximize the total duration of pairings to be covered by the regular crew members during the month.

Gamache and Soumis (1998) present an approach to the two-week rostering problem that is based on column generation embedded in a branch-and-bound algorithm. They showed that the assignment of skeleton activities during optimization is substantially more cost-effective than a pre-assignment.

In the approach of Day and Ryan (1997), first *lines-of-work* are constructed that are consistent with the off-days for each crew member over a sub-roster period. Then an SPP approach is used to determine an optimal feasible sub-roster. These two steps are repeated for each subsequent sub-roster period until a full legal and feasible roster is constructed for the complete roster period.

Dawid et al. (2001) solve the crew rostering problem by using an efficient adaptation of the branch-and-bound technique. The model incorporates several strategies that exploit problem-specific knowledge and rostering-specific properties (e.g. variable selection, branching strategy and cutting-planes) in order to solve even large problems. This approach shortens the solution process and outperforms standard techniques or general-purpose optimizers.

Gamache et al. (1998) present a solution approach for the preferential bidding rostering that uses a sequential method based on seniority order. For each crew member a residual problem is solved. Given a crew member and a set of unassigned pairings, the solution to an integer linear program determines the crew member's maximum-score schedule while taking into account all the remaining crew members. The residual problem is solved by column generation embedded in a branch-and-bound tree. Integer solutions are obtained by using cutting planes.

In general, the crew pairing problem is solved first to reduce the complexity of the crew assignment problem. However, if problem size is small enough, crew scheduling can be solved in one step. An intermediate approach is presented by Chang (2002). In this approach, flight duties are assigned to individual crew members thus bypassing the pairing generation. The approach aims at short-haul problems because in this case the limited time between flights prevents the crew from executing duties in another aircraft. The assignment of duties is much easier than the assignment of the rotations because the length of a duty is a day. The rotation, however, varies from one day to several days. The duty-forming process creates the elementary duties according to the aircraft's schedule and some rules such as the maximum sectors in a duty and crew numbers for duty execution. In order to enhance the efficiency of crew dispatching, the duty assignment process considers all duty connection rules such as minimum rest time and the limitation of flight hours while assigning the duty. Since the assignment unit is a duty, a crew's duty can be changed very easily. Thus, this approach can be used for crew re-scheduling during irregular operations.

2.5 Integrated Models

2.5.1 Overview

The reason for the decomposition of the airline scheduling problem is to reduce complexity. Its subproblems are less complex and easier to solve than the overall airline scheduling problem.

However, if a problem is decomposed, interdependencies between the subproblems are not considered. In addition, in a sequential solution process solutions of previous subproblems limit flexibility of later planning steps (Cordeau et al., 2001). This may result in suboptimal or even infeasible solutions to the overall problem (Barnhart et al., 1998; Cordeau et al., 2001; Mashford & Marksjo, 2001; Klabjan et al., 2002; Barnhart et al., 2003). For example, the fleet assignment can lead to violations of maintenance requirements and increased crew costs, and in flight schedule generation the availability of resources and their costs of assignment are usually not considered (Clarke et al., 1996; Clarke et al., 1997; Langerman & Ehlers, 1997; Barnhart et al., 1998; Kliewer, 2000; Mashford & Marksjo, 2001; Klabjan et al., 2002; Yan & Tseng, 2002; Barnhart et al., 2003; Cohn & Barnhart, 2003).

To cope with these interdependencies, airlines have implemented feedback-loops and iterations between the planning steps in their scheduling environment. Thus, decisions in later planning steps (e.g. crew or aircraft scheduling) are sent to previous steps (e.g. flight scheduling) where manual adjustments and slight modifications are performed (Grandeau et al., 1998; Rexing et al., 2000; Sriram & Haghani, 2003). These iterations are necessary to improve solution quality with this planning paradigm but are at the same time costly and time-consuming.

With advances in optimization theory, algorithms, and computational hardware, researchers were able to solve more complex problems and developed solution approaches to integrate subproblems that were solved separately before (Sherali et al., 2006). In the remainder of this section such solution models are presented. The order of presentation is chosen according to the number of subproblems combined in each approach starting with the approaches that integrate two subproblems of different airline scheduling phases.[20]

2.5.2 Models

Network Design and Frequency Assignment. Many contributions to the flight schedule generation problem integrate the network design and frequency assignment problem, focusing on different aspects such as the airline network structure or the airline hubbing problem (Kanafani, 1981).

In the approach of Berechman and Shy (1996) the choice of a monopoly airline between a hub-and-spoke and a point-to-point network is determined. By introducing flight frequencies and travel time into the passengers' utility function when choosing an itinerary, they find that an airline will choose to operate a hub-and-spoke network if higher flight frequencies can compensate for the inconvenience of having to connect at the hub.

The analysis of Brueckner and Zhang (2001) determines optimal fares and flight frequencies in a hub-and-spoke network and a point-to-point network. The model uses a distribution of desired arrival times across passengers, along with a utility function that depends negatively on schedule delay, which equals the difference between the desired and actual arrival time. The result is that flight frequency is higher in the hub-and-spoke network. Another result is that, despite lower costs per passenger in the hub-and-spoke network, greater flight frequency allows the airline to charge a higher fare to passengers. In this model, a monopoly airline and only fixed flight costs are assumed.

In the model of Ghobrial et al. (1992) an O&D-demand matrix and a set of candidate routes is given. They solve the frequency planning where candidate routes may yield a zero frequency, thus, this approach also partially solves the network design problem.

In the study of Hsu and Wen (2000) a series of models to forecast airline city-pair passenger traffic and to determine the shape of a carrier's airline network and its

---

[20] Sherali et al. (2006) present integrated fleet assignment models and solution approaches. Because this paper does not present a newly developed model, this paper is not included in the following. However, because the authors conduct a thorough analysis of existing approaches, this paper is recommended to get a detailed insight into research on the fleet assignment problem.

corresponding flight frequencies is presented. The exact number of passengers who will connect to another flight at an intermediate airport and the total travel time of individual routes are difficult to estimate because routing and scheduling decisions are later planning steps. To resolve these difficulties, in this study grey clustering to evaluate routes with uncertain and vague parameters is applied. Teodorovic et al. (1994) address this problem using fuzzy logic and single-objective programming; however, their model does not forecast passenger traffic on individual city-pairs.

Lederer and Nambimadom (1998) analyze the choice of a network design and frequency assignment to its route. Their objective is to examine how network choice and scheduling decisions affect airline costs and passenger service quality. The paper demonstrates how structural parameters such as the distance between cities, the demand rates and the number of cities served affect the optimal airline network.

Network Design and Aircraft Routing. Balakrishnan et al. (1990) solve the network design and aircraft routing problem in one step. The objective is to find routes for long-haul aircraft from a main base to one or more terminal bases in order to maximize total profit with respect to given (inter-city) traffic estimates and revenues for each origin-destination pair, aircraft operating costs, and aircraft capacities. The routes of the aircraft then determine the network structure of the airline. The problem is formulated as a multi-commodity network flow problem and is solved using a heuristic procedure based on Lagrangian relaxation. This procedure first finds a feasible solution that is improved with a local search technique afterwards. Because cities need to be indexed in this approach, this solution technique is suitable for carriers with a thin long-haul route network that covers very vast geographic areas in which the distances between intermediate cities are relatively large.

Richardson (1976) optimizes the routing of aircraft of a long-haul international carrier operating in a low density market. The solution to this problem determines the network structure. A mixed integer linear programming approach is formulated that is solved by using Benders' decomposition algorithm. Given demand estimates for each flight leg and operational data, the objective is to maximize profit derived from each routing.

Frequency Assignment and Flight Scheduling. Dobson and Lederer (1993) present a mathematical program to study the competitive choice of flight schedules and route prices by airlines operating in a single-hub system. Assumptions in their model include a single aircraft size, one class of customers, no traffic originating at or destined to the hub airport, airline variable costs dependent on flying time alone and zero variable passenger costs. An additional assumption requires that duopolists serve the identical set of spoke cities using the same hub. The passengers' demand is modeled as a logit function depending on the service and the fare. The service is measured in terms of total travel time and deviation between desired and actual departure time. An airline's total demand is calculated in an iterative two-phase process. In the first phase, a heuristic develops a flight schedule including fares and maximizing profit. In the second phase, competing airlines adjust their schedules and fares. A second heuristic builds an equilibrium regarding frequencies and fares. Both phases are repeated in an iterative process.

Flight Scheduling and Fleet Assignment. Rexing et al. (2000) solve the daily basic fleet assignment problem given a flight schedule and departure time windows. Within each time window the actual departure time may be changed to allow solutions of higher quality. For this, the time windows are discretized into smaller intervals. Because the time windows have a maximum of +/– 10 minutes width, the authors assume that the demand does not vary. A time window of zero width can be assigned to flights that are not allowed to be changed (e.g. for shuttle flights, slots, waves). The problem is modeled as a time-space network. To reduce complexity, a network preprocessing step is conducted.[21]

In the approach of Ioachim et al. (1999), based on a weekly flight schedule with departure time windows, the fleet assignment and flight scheduling problem is solved under consideration of schedule synchronization constraints over the different days of the week. All flights to be performed during a week are scheduled simultaneously instead of solving a daily problem and repeating its solution. The authors use a multi-commodity flow formulation for the fleet assignment problem that includes this new type of constraint. The optimal solution approach is based on the Dantzig-Wolfe decomposition. The problem of schedule synchronization occurs in the long-range planning process of many airlines. When flights with the same identifier are flown on different weekdays, the departure has to be scheduled at the same time every day for marketing purposes. Other requirements, such as maintenance or restrictions on the aircraft types found together at a given base are not included in the proposed problem formulation.

Bélanger et al. (2006) present a fleet assignment problem with time windows. The objective is to schedule the flights within the given time windows so that the flights do not compete with each other. To reduce the effects of competing flights, penalties for flights in the same market departing close to each other and time-dependent revenues are included in a non-linear integer multi-commodity network flow structure which is solved using a branch-and-price strategy.

Flight Scheduling and Aircraft Routing. Levin (1971) proposes the first model for the daily aircraft routing and scheduling problem with variable departure times. Time windows are modeled by allowing departure times to occur at discrete intervals within the time window. The problem is formulated as a network problem with side constraints. To reduce the size of the resulting network, which is huge for wide flight departure time windows, a node aggregation technique is proposed. The authors use an integer linear program with bundle constraints and solve the problem with branch-and-bound methods. Only one fleet is considered.

In the approach of Pollack (1974), departure time windows for each flight as well as already assigned aircraft types are given. His objective is to minimize the fleet

---

[21] This model is used in the integrated solution approach presented in this study in Sect. 4.3. It is presented in detail on pages 100 ff.

size that is necessary to carry out all flights by determining departure times within the given intervals.

Fleet Assignment and Aircraft Routing. Barnhart et al. (1998) present a combined fleet assignment and aircraft routing problem. In an enhanced fleet assignment model, through values and maintenance constraints are considered. The authors introduce flight strings that present sequences of flights that begin and end at maintenance stations. These strings contain the operational costs as well as negative through revenues and opportunity costs for spilled passengers. To solve this problem, the authors use a branch-and-price technique. However, this approach is a slightly modified sequential approach because in the enhanced fleet assignment model only pseudo-maintenance constraints for a sufficient number of aircraft at maintenance stations are considered. Thus, an aircraft routing model has to be solved in the second step to guarantee feasible solutions.

Sriram and Haghani (2003) face the maintenance routing problem including a heterogeneous fleet of aircraft for a seven-day time horizon. The weekly scheduling problem allows the inclusion of less frequent maintenance checks like the B-check. Maintenance is performed at night and flight sequences during the day are supposed to be fixed. In a multi-commodity network flow formulation, the authors suggest a hybrid heuristic of random search and depth-first search to solve this problem.

Fleet Assignment and Crew Pairing. In the approach of Barnhart et al. (1998) the fleet assignment and crew pairing problems are still solved sequentially, but the fleet assignment model is enhanced to incorporate properties of the crew scheduling problem. The relaxation of the crew pairing problem within the fleet assignment model is based on a duty network and ensures that all flight legs are covered by eligible crews. However, constraints on the maximum number of duties within a pairing or the maximum time away from the crew base are not imposed. Because only time-away-from-base is considered as a dominant crew cost, this model may only be suitable for long-haul service where this assumption is reasonable.

Subramanian et al. (1994) formulate the daily fleet assignment problem as a large-scale mixed integer linear program / time-space network in which a combination of operating and passenger spill costs is minimized. The authors propose two different objective functions aiming at minimizing the number of aircraft needed or to maximize profit. During optimization, crew properties are considered by minimizing the number of lonely overnights.

Aircraft Routing and Crew Pairing. The crew pairing approach by Cohn and Barnhart (2003) is extended by incorporating key maintenance routing decisions. With their approach, they can guarantee a crew pairing solution that is also feasible concerning the maintenance constraints.

Based on a given fleet assignment, Cordeau et al. (2001) present an integrated approach with approximate crew costs, in which they link maintenance routing and crew pairing models by a set of additional constraints. The solution process iterates between a master problem that solves the aircraft routing problem, and a subproblem that solves the crew pairing problem. The model incorporates aircraft maintenance constraints as well as the most important crew scheduling constraints. These constraints are in fact the basic ones that must be considered by most airlines and may completely represent the work-rules of a small regional carrier or those of an airline with a simple collective agreement. The authors present a a combination of Benders' decomposition and branch-and-price approach.

Crew Pairing and Crew Assignment. Yan et al. (2002) present a crew pairing model that includes factors such as multiple home bases and mixed aircraft types by building so-called individual pairings (that can be assigned to more than one cabin crew member). To minimize crew costs, the authors present eight models that are formulated as integer linear programs solved by column-generation-based algorithms.

Zeghal and Minoux (2006) present a combined crew pairing and assignment problem for pilots and officers for airlines conducting short and medium haul flights. The problem is formulated as large scale integer linear program in which the decision variables relate to individual crew members (instead of generic crew pairings) with new constraints that replace a large number of binary exclusion constraints. The problem could be solved using standard linear programming technology, however, the authors develop a heuristic that turns out to be more efficient for the test problems.

In the approach of Guo et al. (2006), multiple home bases are included in a partially integrated crew pairing and assignment problem. Before conducting the crew assignment step, this approach first constructs chains of crew pairings taking weekly rests and guaranteed individual pre-scheduled activities into account and including crew base dependent costs and capacities. This problem is formulated as a time-space flow network problem and solved using standard exact optimization techniques. Because of this extension of the crew pairing problem, the following crew assignment can better be solved without additional necessary modifications of the pairing solution. Thus, by considering these elements, the second crew assignment step can be easier solved.

Network Design, Frequency Assignment, and Fleet Assignment. Jaillet et al. (1996) present a model for the problem of route selection, frequency assignment and fleet assignment for hub-and-spoke networks. Because no a priori hub-and-spoke structure is assumed, the proposed network structure is far from looking like a pure hub-and-spoke system, although the design suggests the presence of strong connecting hubs. The authors propose heuristic schemes based on mathematical programming to obtain good solutions. The model is based on the assumption of a single airline operating with a fixed share of market demand. The objective is to minimize the total transportation costs. Inter-city passenger travel demand is estimated based on a simple gravity model.

Wojahn (2002) studies the case of a monopoly airline that maximizes profit. A hub-and-spoke and point-to-point network are compared to each other and a model that allocates the frequency and aircraft capacity to each route is presented.

Adler (2001) attempts to evaluate the most profitable hub-and-spoke network for an airline in a competitive environment. The author determines the airline's routes in a hub-and-spoke network, the frequency of service, aircraft sizes, and air fares.

The competitive model can be defined as a multi-airline, non-cooperative, two-stage game. In the first stage, all competing carriers select (one or two) hubs and connections by using an integer linear programming model that either minimizes the great circle distance between hubs and spokes or the total number of passengers flying on more than one-leg journeys. Once connections are set, they cannot be changed within the game. An airline can choose not to connect a spoke through low frequency, but the spoke cannot be attached to a different hub once the first stage has been completed. The second stage of the game pits competing carrier networks against each other based on frequency, aircraft sizes and air fares based on the knowledge of all the airlines' choices during the first stage. Decision variables are computed using a multi-nomial logit market share model implanted in a nonlinear mathematical program which computes airline profits (or losses).

Flight Scheduling, Fleet Assignment, and Aircraft Routing. Soumis et al. (1980) describe a model that iteratively solves the aircraft routing problem of selecting the flights that improve profits, and the problem of passenger assignment to these flights to estimate revenues. The objective is in general to minimize the difference between the utilization cost of the aircraft of different fleets and passenger revenue, taking into account the loss of revenue associated with passenger dissatisfaction caused by the quality of service. A partial heuristic optimizes the schedule by adding or dropping flights if there is an improvement in the objective function. This model is on a daily basis and does not provide recapture possibilities, thus if the demand is larger than the capacity of one aircraft route, revenue is lost. It is also assumed that the airline can specify a set of candidate flights probably including most of those in operation.

Desaulniers et al. (1997) solve a daily aircraft routing and scheduling problem. They provide a fleet type for each leg, the routing for each aircraft, and departure times given different fleets (including their operational and cost data), flight legs over a one-day horizon, and departure windows. As formulation for this problem, the authors present two models. One is an SPP-type formulation and another one is a time constrained multi-commodity network flow formulation. A column generation technique is used to solve the linear relaxation of the first model and a Dantzig-Wolfe decomposition approach is used to solve the linear relaxation of the second model.

Flight Scheduling, Aircraft Routing, and Crew Pairing. Klabjan et al. (2002) reverse the order in which they consider the maintenance routing and crew pairing problems. To achieve feasible solutions of the aircraft routing problem, aircraft count constraints are added to the crew scheduling model. However, this model does not guarantee maintenance feasibility. In addition, this model allows the departure time of each flight leg to be moved within a given time window to further reduce crew costs. The flexibility in departure times should result in feasible pairings that would have been infeasible based on the original schedule.

Fleet Assignment, Aircraft Routing, and Crew Pairing. Clarke et al. (1996) solve the fleet assignment problem under consideration of maintenance and crew constraints. Thus, the fleet assignment model accounts some of the downstream effects on later planning steps but does not include these elements as integral component of the optimization. While retaining solvability, better crew connections are constructed. Although crew costs are not explicitly considered, they can be reduced by avoiding lonely overnights. Aggregate maintenance constraints are included that require a minimum number of maintenance opportunities. These aggregate constraints are approximate because they do not guarantee that the maintenance opportunities are distributed equally among the individual aircraft. One aircraft may have more maintenance opportunities than it needs whereas another may have none. Thus, this approach still may not yield a feasible solution to the aircraft routing problem because individual aircraft are not considered.

Rushmeier and Kontogiorgis (1997) solve the daily fleet assignment problem. They use an enhanced multi-commodity flow network which represents connection time rules in their full complexity. This allows the recognition and treatment of connection possibilities to utilize aircraft most efficiently. To capture interrelations with later planning steps, crew and maintenance considerations are integrated in the objective function as soft constraints. The model does not consider the full aircraft routing implications.

Network Design, Frequency Assignment, Flight Scheduling, and Aircraft Routing. Erdmann et al. (2001) address a problem consisting of the flight schedule generation and aircraft routing problem for a charter airline. The problem is a special case of the schedule generation problem which particularly arises in charter business. Based on a given fleet and a set of origin-destination pairs with associated passenger demands (evenly distributed), aircraft rotations for all aircraft of the fleet are created that maximize profit. Departure and arrival times are introduced as much as necessary to model the passengers' switching of aircraft. The problem is modeled as a capacitated network design problem and solved by a branch-and-cut approach.

In the approach of Mashford and Marksjo (2001) a model to optimize generic airline routes using a simulated annealing approach (SA) is developed. Each solution is represented as a flight graph. During a simulated annealing run, current solutions are modified in two ways: The local substitution operator acts by randomly choosing an aircraft, choosing a sub-tour of the tour of that aircraft and then substituting an allowable sub-tour for the chosen sub-tour by either (randomly) changing flight legs in an aircraft's route or by changing departure times of flights. The timing adjustment operator acts by randomly selecting a departure and then randomly changing the departure time in such a way as to maintain feasibility. In the evaluation phase necessary in each iteration of the SA, the current schedule is evaluated with regard to operating profit. Two main components in this step are the connection builder and the passenger assignment builder. Whereas the connection builder constructs feasible and desirable O&D-itineraries, the latter model assigns passengers to these itineraries using a calculated attractiveness by simulation. The reason for this is the modeling of the actual behavior of an individual passenger and the interaction with a reservation system. The actual behavior of a passenger leading to a reservation is a function of the reservation system state, thus, the reservation system state evolves with time.

Network Design, Frequency Assignment, Flight Scheduling, and Fleet Assignment. The model of Lohatepanont and Barnhart (2004) integrates the schedule design and the fleet assignment. The schedule design part of the complete procedure selects flights to be included into the schedule from a given set of candidate flights. Thus, this part includes the subproblems network design, frequency assignment, and flight scheduling. The set of candidate flights consists of two groups of flights: a mandatory flight list with flights that must be included into the schedule, and an optional flight list that the algorithm may, but need not, include into the schedule. The selection of flights from the complete set and the assignment of fleet types to these flights is simultaneously optimized by extending the fleet assignment model from Barnhart et al. (2002) with demand correction terms (see page 23). These variables modify the input demand for the model as changes to the schedule are made. Their calculation is based on a schedule evaluation procedure that estimates the demand for each itinerary. As it would be necessary to perform an exponential number of evaluation runs (one run for each possible set of flights) to determine the correction terms, rough estimates are used and iteratively revised as needed. Another approach that does not continuously adjust demand as the schedule changes is an approximation of the interactions between demand and supply using recapture rates which are iteratively revised.

Network Design, Frequency Assignment, Flight Scheduling, Fleet Assignment, and Aircraft Routing. Yan and Tseng (2002) present an integrated model for the flight schedule generation and aircraft scheduling problem. Neither draft timetable nor preset operational flight leg information are utilized in the proposed model. Instead, more accurately given trip demands and all the supply constraints (e.g. aircraft types, fleet size and airport slots, airport quotas) and related cost data serve directly as the model's basic input. The objective is to construct profit-maximizing (daily) aircraft rotations. Maintenance or through flight considerations are not included in the aircraft routing subproblem and departure intervals for flights are given. Demand for flights only depends on fares. The problem is formulated as an integer multi-commodity network flow problem with side constraints. An algorithm based on Lagrangian relaxation, a sub-gradient method, the network simplex method, the least cost flow augmenting algorithm and the flow decomposition algorithm are developed to solve the problem.

2.6 Summary, Conclusion, and Future Challenges

2.6.1 Summary

An airline schedule represents the central element within an airline's corporate planning framework and consists of two elements:

- the flight schedule that contains detailed information on the offered flights (like departure and arrival times and airports) and that is presented to potential passengers, and
- the assignment of aircraft and crew resources to the flights.

The airline schedule is an airline's primary marketing tool, having the largest influence on passenger demand. Furthermore, it affects every operational decision. Given an airline schedule, a significant portion of revenues and costs is fixed. Consequently, airline scheduling is one of the most important planning tasks in an airline. However, because of the large number of factors that have to be taken into account simultaneously when constructing or optimizing an airline schedule, this problem also represents the most demanding challenge for the airline.

Because of its complexity, the consensus today is that a single model for the airline scheduling problem is computationally intractable and that this planning problem is best solved in a sequential process. The overall problem is decomposed into subproblems of less complexity that are solved in a sequence. Subproblems are then aggregated to phases in the airline scheduling process. One possible decomposition and solution process can be found on page 10, consisting of the phases flight schedule generation, aircraft scheduling, and crew scheduling. The purpose of this chapter was to introduce these subproblems including their objectives and major constraints as well as presenting optimization models that were developed to solve these subproblems.

Solution approaches to more than one subproblem were presented in Sect. 2.5. Many researchers developed solution approaches that integrate two or more subproblems of the airline scheduling problem and reported (significant) improvements in solution quality. In some models, a second subproblem is only approximated. For example, if in a fleet assignment model maintenance considerations are included by ensuring sufficient maintenance opportunities, the aircraft routing problem might still be infeasible because the spacing of maintenance visits and individual aircraft were not considered (Clarke et al., 1996; Clarke et al., 1997). One interesting result of the previous section is that there is only little work integrating crew scheduling problems to the flight schedule generation and aircraft scheduling phase. Although there might be a high potential for increases in profits because of high crew costs, in this study it is believed that the following three facts might be the main reasons for this observation:

- Crew work-rules are highly affected by company policies and legal restrictions, thus, these rules may differ from airline to airline and country to country, as do many crew costs. Because of this, an optimization approach that solves a generic crew scheduling problem might be of theoretical interest only.
- Because of the high number and variety of restrictions, the crew scheduling problem is a very complex problem by itself, and, thus, it is even harder to solve a combination of this problem and another airline scheduling problem.
- To solve the crew scheduling problem, human factors like team quality and satisfaction have to be considered that are hard to quantify. Thus, many crew

scheduling problems are still solved manually or using simple heuristics embedded in systems that support key decisions of human schedulers.

2.6.2 Conclusion

Optimal solutions of the airline scheduling problem can only be realized if all relevant variables, their interdependencies, and restrictions are combined in one model of considerable detail, and a solution algorithm that guarantees to find the optimal solution is applied. Optimal solutions to subproblems do not imply an optimal overall solution. When decomposing a problem, interdependencies between subproblems cannot be considered. Solving subproblems in a sequence will result in less freedom for later planning steps, because the solution of one planning step serves as given input for the succeeding problem. Thus, solutions are unsatisfactory or even infeasible (Barnhart et al., 1998; Cordeau et al., 2001; Mashford & Marksjo, 2001; Yan & Tseng, 2002; Cohn & Barnhart, 2003; Barnhart et al., 2003; Guo et al., 2006). To overcome these problems, airlines have to implement (costly and time consuming) feedback-loops or iterations in their airline scheduling process (Grandeau et al., 1998; Andersson et al., 1998). Information of later planning steps are propagated to previous subproblems to alter their solutions. To relax the boundaries between the individual solution steps, models have to be developed that include decision variables of more subproblems. This trend towards an integrated airline scheduling model can be recognized in recent publications.

In the beginning of modeling the airline scheduling problem, computational power was limited and research was focused on single subproblems with simplifying assumptions (Desaulniers et al., 1997; Sriram & Haghani, 2003). Advances in optimization theory and computer hardware then led to (Yu, 1998; Barnhart et al., 2003)

- the application of exact solution approaches and, thus, a higher solution quality (Rushmeier & Kontogiorgis, 1997),
- more realistic models, as a problem could be formulated more detailed with a higher number of practical requirements, with less simplifications, and for more realistic problem sizes,
- the extension of the scope of problems by integrating different subproblems (or elements) of the airline scheduling problem.

In general, each airline scheduling model represents a trade-off between these three directions. If more complex problems are considered either by increasing the level of detail or by extending the scope, usually heuristics or exact approaches including heuristic elements are applied. For example, many heuristics are used to solve crew scheduling problems or to enhance exact approaches in this scheduling phase. Some problems are still solved manually with little optimization because either no models exist or they contain major simplifications that lead to an unrealistic problem formulation. For example, most of the presented models for flight schedule generation do not consider relevant restrictions or costs of airline resources, and if so, they represent this information only on a very rough and unrealistic basis. Thus, the flight schedule generation phase is still performed manually with much subjective judgment and decision making. There is little optimization in managing the interrelation between supply and demand directly, systematically, and accurately (Yan & Tseng, 2002).

2.6.3 Future Challenges

The models presented in this study differ in the amount, variety and accuracy of simplifying assumptions. One common characteristic to almost every model is the postulation of deterministic data like fixed demand (or its distribution), flight times, and turn times. However, in reality many of the inputs to the airline scheduling problem are of stochastic nature (Day & Ryan, 1997). The objective of minimizing costs and maximizing revenue in deterministic models usually leads to a very tight schedule with very short turn times (Langerman & Ehlers, 1997). In this case, stochastic deviations in the planned times will result in system-wide schedule delays. Research concerning the relationship between airline market shares and schedule punctuality showed the significance of passengers switching between airlines, once they experience unsatisfactory services from an airline (Caulkins et al., 1993; Suzuki, 2000). This has led to several approaches that aim at developing robust airline schedules that are less susceptible to flight delays (Sherali et al., 2006). The proper choice of schedule buffer time for turnarounds increases the reliability of flight connections at airports (Wu & Caves, 2000; Wu, 2005; Lan et al., 2006). Thus, a quality measure for airline schedules should include profit and its performance in operations.

Until now, the airline scheduling problem has been considered as an isolated planning problem within the airline corporate planning system. This development is supported by the schedule's central role and its major effect on revenues and costs of the airline. However, a second problem in airline planning that has attracted many researchers is *revenue and yield management*. Its objective is to maximize revenue by selling as many tickets as possible at the highest price possible.[22] Since this problem uses a given airline schedule as input, there is a potential for higher profits if revenue management and scheduling issues are solved simultaneously (Jacobs et al., 2000; Barnhart et al., 2003).

Until now, the flight schedule generation phase has attracted only little attention for optimization models, mainly because of its large complexity. The models that are built to support this phase usually do not consider the availability of resources or the costs and implications of their assignment, and if so, the level of detail is much too low (Yan & Tseng, 2002). Other optimization models incorporating flight schedule generation issues mostly adjust departure times of flights within given time windows. In addition and not limited to the flight schedule generation phase, many solution models are rather simplified, disregarding many practical requirements, and

---

[22] More details of this topic can be found for example in Belobaba (1987), Kimes (1989), Smith et al. (1992), Vinod (1995), Weatherford (1998), McGill and Van Ryzin (1999), Subramanian et al. (1999), Belobaba and Farkas (1999), Pak and Piersma (2002), Barnhart et al. (2003), Cote et al. (2003), Pulugurtha and Nambisan (2003).

include assumptions that do not represent reality (for example, a monopoly situation, uniformly distributed demand, only one fleet type, a pure (one) hub-and-spoke network, no maintenance capacity constraints etc.). Thus, there is a need for optimization models of sufficient detail and scope that capture the critical interactions among the various resources of the airline, its competitors, and airports to support this airline scheduling phase (Yan & Tseng, 2002; Barnhart et al., 2003).

The number of subproblems integrated in one model and the intensity of integration needs to be improved in order to achieve solutions of higher quality (Barnhart et al., 2003; Sherali et al., 2006). Some researchers present integrated models that incorporate some elements of a second problem (major costs or constraints), an iterative approach, or an enhanced / advanced but still sequential procedure. Although these models produce (far) better results, regardless of any enhancements, it is believed that no kind of sequential or decomposed solution approach will produce better or equal solutions to an integrated approach. Only an integrated or simultaneous approach including all subproblems could provide a feasible and optimal solution to the airline scheduling problem (Barnhart & Talluri, 1997; Barnhart et al., 1998; Cordeau et al., 2001; Klabjan et al., 2002; Barnhart et al., 2003; Cohn & Barnhart, 2003).

To summarize, according to the directions of research efforts, future challenges in airline scheduling can be outlined according to the following objectives (Barnhart et al., 2003):

- improvement of solution quality by reducing randomness in solution approaches,
- incorporating stochastic and uncertain elements in the scheduling process to increase the robustness of the resulting schedules,
- combination of the airline scheduling with revenue management,
- extending the applicability of optimization methods to a larger number of subproblems of the complete airline scheduling problem (like flight schedule generation),
- representing airline operations at a higher level of detail, thus, reducing simplifying assumptions and including practical restrictions, and
- relaxing the boundaries between the subproblem in the planning process towards an integrated approach.

All these challenges are of major importance to further improve the process of airline scheduling and to obtain realistic schedules that are optimal regarding the airline's overall success. However, the research presented in this study focuses on the last three directions, with the last challenge representing one of the most formulated objectives in airline scheduling research.

Chapter 3
Foundations of Metaheuristics

Abstract. Many real-world problems with practical importance are large and complex, differing from standard problems. Often they belong to the class NP making them computationally intractable using exact optimization algorithms. Metaheurstics are general-purpose improvement heuristics that are used to solve NP-hard problems. Some problem-specific adaptions of the metaheuristic are necessary to achieve an efficient search process. This problem customization focuses on four basic design elements that every metaheuristic incorporates and that are presented in detail in this chapter: the combination of representation and search operators, the fitness function, the initialization, and the search strategy. With respect to the search strategy, in general the different variants of metaheuristics can be classified into two groups: techniques based on local search and techniques using recombination-based search operators. For each search concept, one metaheuristic was specified as a representative example: threshold accepting as a local search routine and genetic algorithms as a recombination-based search. They are the underlying techniques of the simultaneous airline scheduling process presented in this work.

3.1   Introduction

To reach a certain pre-defined goal, a decision maker has to choose between different decision alternatives. Planning describes this process of generating and comparing different courses of action and then choosing one of them prior to action (Rothlauf, 2006a). As each alternative might have a different impact on the decision maker's objective, it is necessary to evaluate each given alternative with regard to reaching the goal and to assign a quality value. Furthermore, restrictions and limitations have to be taken into account that limit the freedom of action. One possible decision alternative is denoted as the solution of the problem, the complete set of possible solutions establishes the search space of the problem. In optimization problems, the objective is to find the best solution: the decision alternative with the highest contribution to the overall goal.

T. Grosche: Computational Intel. in Integrated Airline Scheduling, SCI 173, pp. 47–57.
springerlink.com

Planning processes and the solution of optimization problems are the core discipline of operations research (OR) (Taha, 2002; Hillier & Lieberman, 2002; Domschke & Drexl, 2005). Its focus is on model construction and model solution:

- **Model Construction:** The objective of this step is to simplify the given real-world problem to make it computationally tractable. The properties of the real-world problem, its influencing factors, and relationships are represented as a mathematical model consisting of a set of symbols and expressions. Decision variables represent the properties of the different decision alternatives. The quality of an alternative is calculated using an objective function. Any restrictions on the problem or its variables are expressed by a set of constraints.
- **Model Solution:** Given an optimization model, algorithms are applied to solve the model. Because many different solution techniques were developed in the past, this step is straightforward if an appropriate model was constructed, requiring only the choice of one suitable solution method. In fact, OR focuses on model construction which itself is demanding and requires the researcher's experience. Sometimes, model solution is considered only as a (simple) subsequent step (Ackhoff, 1973).

Many of the standard optimization algorithms developed by OR are exact techniques that guarantee to find the optimal solution of the given problem. If their effort grows polynomially with the problem size (P-problems), exact optimization algorithms can be applied. However, many problems belong to the class NP. Problems of this class are less structured than P-problems (for example, the objective function is not differentiable, the objective function and constraints and variables are not linear and continuous etc. (Rudolph & Schwefel, 1994)). Then, enumeration techniques are the only exact algorithms that can be used to solve these problems. As the effort for problem solution grows exponentially with problem size, these techniques can only be applied if the problem size is low enough.

3.2 Metaheuristic Optimization

Many real-world problems with practical importance are large and complex, differing from standard problems. Often they belong to the class NP making them computationally intractable using exact optimization algorithms. To solve these kinds of problems, heuristic optimization techniques could be used. Heuristics often exploit properties of the problem and use knowledge about high-quality solutions or rules of thumb when searching for good solutions (Pearl, 1984). They often have lower computation times than exact techniques allowing their application to complex problems that are realistic and near to real-world problems. However, they do not guarantee to find the optimal solution. In general, there are two kinds of heuristics: construction heuristics develop a solution to a problem by iteratively adding elements to partial solutions until a complete solution is achieved, whereas improvement heuristics start with a complete solution and iteratively improve the solution by slight modifications.

Since heuristics are problem-specific, properties of the problem to be solved need to be incorporated into the design of the heuristic. If the heuristic does not fit the problem, its solution is likely to be not possible. Thus, the development of heuristic optimization techniques is a difficult process that needs to be performed very carefully. In contrast to heuristics, metaheuristic optimization techniques represent search strategies that are problem independent and, thus, widely applicable (Glover, 1986). Often these techniques are inspired by search strategies from other domains (e.g. biology, physics etc.) and a huge variety of different types of metaheuristics with slightly different properties and functionalities were developed (Bäck et al., 1997; Glover & Kochenberger, 2003). In general, metaheuristics represent extended variants of improvement heuristics (Reeves, 1993; Rayward-Smith et al., 1996; Rayward-Smith, 1998; Michalewicz & Fogel, 2000; Rothlauf, 2006a).

In general, the application of a metaheuristic is straightforward, since only two conditions must be fulfilled (Rothlauf, 2006a):

- solutions of the problem have to be represented as a set (or string) of variables or symbols to allow its processing, and
- the quality of each solution must be quantifiable to allow pairwise fitness comparisons.

Although the basic search strategy of a metaheuristic is independent of the problem, leading to many different applications in various domains, there still is a conflict between ease of application and effectiveness.[1] The broader the applicability of an optimization technique, the poorer are the results of these techniques on individual problems. This relationship is expressed by the No-Free-Lunch-Theorem (NFL) published by Wolpert and Macready (1997). In short, the NFL states that no heuristic can outperform other optimization methods if they do not inherit problem-specific knowledge. Thus, traditional heuristics exploiting problem-specific information might outperform standard metaheuristics, and there is a need to consider problem-specific knowledge in the design of metaheuristic optimization techniques to achieve superior performance (Droste & Wiesmann, 2002; Puchta & Gottlieb, 2002; Bonissone et al., 2006; Rothlauf, 2006a).

3.3 Design Elements of Metaheuristics

In his research, Rothlauf (2006a) examines the design of metaheuristics, the influence of different design variants and the consideration of problem-specific knowledge on their performance. Four basic design elements are identified that all metaheuristics have in common and that characterize the different variants:

1. solution representation and variation operators,
2. fitness function,
3. initialization, and
4. search strategy.

---

[1] For overviews see for example Biethahn and Nissen (1995), Osman and Laporte (1996), Bäck et al. (1997), Alander (2000), Blum and Roli (2003).

In the following, a short introduction to each element is given. Each element is subject of research on metaheuristics and many studies are available that can be used for further information.[2]

3.3.1 *Solution Representation and Variation Operators*

The objective of variation or search operators is to modify solutions of a problem during an optimization procedure. For this purpose, it is necessary to encode any given solution of a problem as a string that can be processed by these operators. The solution of a problem is denoted as the phenotype, its representation as the genotype, the mapping between the phenotype and the genotype is referred to as the representation of a metaheuristic (Rothlauf, 2006b). Because the search operators work on the genotypes, the representation and the operators cannot be treated independently but must be considered as a joint element of a metaheuristic.

Although not selective, direct and indirect representations can be distinguished. When using a direct representation, an explicit mapping is not specified. Instead, solutions are represented in their most natural search space and the variation operators are applied directly to the solutions. This calls for individual variation operators, since the solution representation is problem-specific. In contrast, an indirect representation has an explicit mapping: problem solutions are represented as standard data structures (for example binaries, integers etc.). This allows the application of standard search operators to the genotypes. An additional advantage of using indirect representations is that constraints or restrictions in the search space may be efficiently modeled by a specific encoding (only feasible solutions can be processed) or that an advantageous mapping might decrease problem difficulty (Rothlauf, 2006a). However, finding a proper representation then is a challenge in metaheuristic design.

Search operators can be classified as local search operators and recombining search operators. In the search space, a local search step moves from one solution to a solution in its neighborhood. The concept of neighborhood aims at similarities between solutions. If there is a way to quantify the similarities or common elements, a neighborhood of a solution can be defined that contains all solutions that are similar to the current solution to a specific extent (distance). A high locality exists if this distance corresponds to the distance between their objective values and the distance of the genotypes. If the genotype encodes continuous values, the real difference between the values can be used as distance. If there are discrete values (for example discrete choices), the quantification becomes difficult. However, the number of decision variables encoded and subject to variation allows a quantification of the differences and, thus, allows the definition of a distance and the neighborhood. The local search operator moves to a neighboring solution by modifying the current solution, thus, in fact, the distance of the genotype is defined by the search operator: each solution that can be obtained by a single search step is a neighboring solution.

---

[2] See for example Holland (1975), Grefenstette (1985), Michalewicz and Fogel (1989), Goldberg et al. (1989), Liepins and Vose (1990), Storer et al. (1995), Michalewicz and Schoenauer (1996), Bäck et al. (1997), Goldberg (2002), Rothlauf (2006b), Rothlauf (2006a) etc.

When using local search operators, it is assumed that the structure of the search space can guide the search because good solutions are grouped together. Then, the optimal solution can be obtained by applying only small changes to a solution and the objective value of solutions considered earlier is used to guide the future search process (Manderick et al., 1991). Local search operators can only yield optimal solutions if the search steps are sufficiently small and if the global optimum can be reached by neighboring steps. If multiple optima exist, there is the chance that a local search operator only yields a local optimum instead of the global optimum. In contrast to local search operators, recombination operators require at least two solutions, because these operators recombine elements of solutions to construct one or more new solutions. The original solutions are commonly referred to as parents, the new solution as child or offspring. When applying recombination operators, it is assumed that the problem under investigation is decomposable: subproblems can be solved independently and the combination of the optimal partial solutions yields an overall optimal solution. Recombination operators represent global search operators, because the resulting solution inherits properties of both parents and is not limited to the neighborhood of one parent. General design principles for recombination search operators were formulated by Radcliffe (1991) and (1994).

When choosing a combination of representation and operators, the bias and locality have to be taken into account. A bias exists if the choice of representation/operators alone pushes the search of the metaheuristic into a specific direction (Caruana & Schaffer, 1988). For example, the operators work not randomly but perform only selected modifications to the solutions. High locality is necessary to allow a guided search. If the locality is high, small changes to the genotype by a variation operator result in small changes of the phenotype and its objective value (Lohmann, 1993; Rothlauf, 2006b). Thus, with each application of the operator, a neighboring solution in the search space is obtained. If the locality is low, the search process represents a random walk through the search space.

3.3.2 Fitness Function

A fitness function assigns a fitness value to each solution, which is used by the metaheuristic to compare the quality of solutions. In many cases, the fitness function corresponds to the objective function of the problem behind it. This objective function expresses the quality of the current solution with respect to the goal that is to be achieved. In general, this is the primary objective of the optimization process. However, sometimes the fitness or objective value is modified by the metaheuristic to improve the metaheuristic's performance or to include additional characteristics of the corresponding solution. One example for performance improvements is to smooth or scale the fitness landscape for guided local search techniques. Beside any additional objectives, a very important characteristic of a solution that can be incorporated in the fitness value is violations of given restrictions. A solution might be infeasible but could inherit some favorable properties that should be kept in the optimization process. Thus, instead of removing the current solution from the search process, its fitness value can be decreased by a penalty function.

Metaheuristics represent iterative search procedures, and usually a large number of fitness evaluations is necessary. Thus, their application is only possible if the fitness calculation is not too complex and does not require too high an amount of computational time and effort. On the other hand, metaheuristics can work only with pairwise fitness comparisons, and especially in the beginning of the optimization process, large differences in the fitness value of different solutions exist. Thus, if necessary, fitness values can either be obtained by approximations (at the beginning of the optimization process) or can focus only on the differences of similar solutions (at the end of the optimization process).

3.3.3 *Initialization*

In the initialization step, a solution is presented to the metaheuristic and serves as an initial solution from where the optimization process starts. If the metaheuristic uses recombination operators, the initialization must provide a population of solutions to allow recombination between different solutions. If some problem-specific knowledge or information about high-quality solutions exists, the initial solutions can be constructed using this knowledge to lead to a good starting point for the optimization process. If there is no such knowledge, the initial solution has to be created randomly. All possible solutions then have to have the same selection probability and no specific solution or solution properties are favored. If a population needs to be initialized, the diversity of this population should be high to ensure an effective application of the recombination search operator.

3.3.4 *Search Strategy*

Decisions on the search strategy of a metaheuristic focus on the exploration and exploiting phases (Blum & Roli, 2003; Rothlauf, 2006a). In exploitation, the search is focused on promising regions in the search space, whereas in exploration new areas in the search space are investigated. In general, two basic search strategies can be identified. They depend on the main search operator used during optimization, leading to local search and recombination-based search strategies. The control of exploitation and exploration, or intensification and diversification, is addressed differently in each strategy, and variants exist each focusing on different mechanisms for intensification and diversification (see Rothlauf (2006a) for a selection of representative examples).

In local search strategies, a new solution is iteratively chosen from the neighborhood of the current solution. The fitness of the solution is used to guide the search process to regions of the search space with high-quality solutions (intensification). To escape from local optima, diversification is necessary. This can be accomplished by further varying the solution for example by changing the definition of a neighborhood of a solution. An alternative would be to start multiple instances of the optimization process, each with a different initial solution. A very common strategy to allow diversification is not to restrict search steps to intensification (improvement) steps but to allow inferior solutions during the search. Then such search steps without improvement are the result of random and larger steps or are based on information of the search space from previous steps.

In contrast to local search strategies, recombination-based strategies already inherit diversification because the initial population consists of different solutions with different properties. During the application of recombination-based search operators and selection, which removes low-quality solutions from the population, new solutions are created with similar properties to the existing solutions, thus, diversification is reduced. To prevent the situation in which all diversity is lost and the population has converged to a non-optimal solution (premature convergence), diversification strategies have to be applied. Besides strategies from local search, these strategies include the increase of the number of solutions in the population, the reduction of the selection pressure (thus, accepting inferior solutions), the limitation of the number of similar solutions, or the processing of different sub-populations simultaneously.

3.4 Selected Metaheuristic Optimization Techniques

In the following, two metaheuristic optimization techniques are presented, each chosen as a simple and representative example for local and recombination-based search. They are the underlying techniques of the simultaneous airline scheduling process presented in Sect. 4.4.

3.4.1 Local Search: Threshold Accepting

Threshold Accepting (TA) was developed by Dueck and Scheuer (1990) and represents a formally very similar method to simulated annealing (SA) developed by Kirkpatrick et al. (1983) and Cerny (1985). SA has become a popular general purpose tool for a wide class of combinatorial optimization problems. However, Dueck and Scheuer (1990) show that for many problems TA achieves at least the same solution quality as SA while requiring considerably less effort.

TA represents an iterative local search that is capable of escaping from local optima by accepting inferior solutions during optimization. This diversification element is provided by a random walk. In each search step, a variation operator modifies the current solution s. The resulting neighboring solution s^* is evaluated using the fitness function f. If the fitness value $f(s^*)$ of s^* is higher than $f(s)$, s^* replaces the old solution s. If $f(s^*)$ is lower than $f(s)$, the new solution is only accepted if the fitness loss $\Delta f = f(s) - f(s^*)$ is less or equal to a given threshold T. This allows an escape from local optima. T represents the strategy parameter of TA and is set to a sufficiently high value to allow high diversification at the start of the optimization process. During optimization, T is reduced until $T = 0$. Then, TA represents a local search in which only better solutions are allowed to replace the current solution. At this point, only intensification is existent, there is no further diversification in the search process.

When applying TA to a problem, the researcher has to carefully select the initial threshold and the reduction schedule of the threshold. There is no standard procedure for this task and many different options exist to decide on the threshold and its reduction. Often, the threshold is reduced by a certain value after a given number of iterations. In the following algorithm 1, one example for the basic functionality of TA is presented.

Algorithm 1. Threshold Accepting

1: choose initial threshold $T > 0$

2: choose threshold reduction step size r

3: choose maximum number i_{max} of iterations between improvements

4: choose maximum number t_{max} of iterations between threshold reduction

5: create initial solution s with fitness value $f(s)$

6: $t = 0$

7: **repeat**

8:    $t = t + 1$

9:    create neighboring solution s^*

10:    calculate new fitness value $f(s^*)$

11:    $\Delta f = f(s) - f(s^*)$

12:    **if** $\Delta f < T$ **then**

13:       $s = s^*$

14:       $i = 0$

15:    **else**

16:       $i = i + 1$

17:    **end if**

18:    **if** $i > i_{max}$ or $t > t_{max}$ **then**

19:       $T = T - r$

20:       $i = 0$

21:       $t = 0$

22:    **end if**

23: **until** termination

3.4.2 *Recombination-Based Search: Genetic Algorithms*

Genetic algorithms (GA) are recombination-based metaheuristics and belong to the class of evolutionary algorithms (EA). EA were introduced by Holland (1975) and Rechenberg (1973b) and have been applied to a variety of different problems of different domains.[3] These optimization techniques are inspired by evolutionary

---

[3] Some applications of EA to airline-related problems can be found for example in Levine (1996), Langerman and Ehlers (1997), Christou et al. (1999), Gu and Chung (1999), Ozdemir and Mohan (1999), Ozdemir and Mohan (2001), Pulugurtha and Nambisan (2001b), Pulugurtha and Nambisan (2001a), Chang (2002), Pulugurtha and Nambisan (2003), Caprì and Ignaccolo (2004), Lee et al. (2007).

principles and imitate basic biological operators of the modern evolutionary synthesis. This theory of evolution is based on the findings of Darwin (1859) and Mendel (1866) and identifies selection, recombination, and mutation as the basic mechanisms of nature to propagate advantageous properties of creatures throughout populations (*survival of the fittest*). In EA, these mechanisms are formulated at an abstract level to solve optimization problems.

Many different variants of EA have been developed since its beginnings. They differ in the design elements presented in Sect. 3.3 and the emphasis of certain operators or their intended use (for example mathematical optimization vs. machine learning). However, each technique relies on the basic evolutionary principles and the differences between the different variants have become much less in more recent algorithms, especially when applied to real-world problems. Besides GA, two major kinds of EA can be identified and are subject to current research: evolution strategies (ES) and genetic programming (GP). Basic information regarding these techniques can be found for example in Rechenberg (1973a) and Rechenberg (1973b) for ES and Koza (1992) for GP.

3.4.2.1 Simple Genetic Algorithm

The simple GA denotes a GA in a basic form and, thus, well describes the functionality of a GA (Goldberg, 1989). The algorithm uses a population S of n solutions s during the search process. A solution is denoted as individual and usually consists of a string of fixed length l incorporating the problem parameters. New solutions (a new generation) are created by applying recombination-based operators (*crossover* or *recombination*) to the existing solutions. Usually, a crossover operator creates a new offspring of two parental solutions by exchanging substrings of the parents. The crossover represents the main variation mechanism of the simple GA. In addition, a local search (*mutation*) is performed on individual solutions, that serves as a background operator. This operator has the additional functionality to insert new properties into the solution, that might not be existent in the current population. A selection operator decides which solutions are removed from the population and are no longer available to the search operators. The decision on the removal of solutions might be controlled deterministically or stochastically. On average, low-quality solutions have to be removed from the population to guide the search towards promising regions in the solution space. All operators are applied iteratively, each iteration is denoted as generation.

The following algorithm 2 presents one example for the basic functionality of the simple GA including its parameters.

Often, a maximum number of generations or the convergence of the population is used as a criterion for termination. A convergence criterion could be the difference between the fitness of the best solution in the population and the average fitness of the population (or the worst solution in the population): the smaller this difference, the higher the convergence. Another option would be to stop the GA if the effort for obtaining new solutions is more expensive than the possible fitness gain (Wendt, 1995).

Algorithm 2. Simple Genetic Algorithm

1: choose population size n

2: choose recombination probability p_r and mutation probability p_m

3: create initial population S with n solutions $s = (s_0, \ldots, s_l)$

4: calculate fitness value $f(s)$ for each $s \in S$

5: **repeat**

6:      insert n solutions from S into a mating pool M using a selection scheme depending on fitness values

7:      $S^* = \{\}$

8:      **while** $|S^*| < n$ **do**

9:        **if** $random(0,1) < p_r$ **then**

10:           create a new solution s^* by recombination of two randomly chosen $s \in M$

11:           include s^* in S^*

12:       **else**

13:           copy one randomly chosen $s \in M$ as s^* in S^*

14:       **end if**

15:     **end while**

16:     **for all** $s \in S^*$ **do**

17:       **for** $i = 0$ to $i = l$ **do**

18:         **if** $random(0,1) < p_m$ **then**

19:             mutate s_i

20:         **end if**

21:       **end for**

22:     **end for**

23:     calculate fitness values $f(s)$ for all $s \in S^*$

24:     $S = S^*$

25: **until** termination

3.4.2.2   Steady-State Genetic Algorithm

The simple GA represents a generational GA. A new generation of solutions is created by applying genetic operators to the current generation until enough new solutions are obtained. Thus, there is an explicit distinction between the parental and offspring generation. One drawback of this approach is the risk that high-quality solutions get lost if their offspring have a lower fitness value than themselves (Reeves & Rowe, 2003). To prevent these situations, concepts like elitism and population overlaps were developed (De Jong, 1975). In elitism, the best individual is not allowed to be replaced. Population overlaps describe the situation when only a fraction (the generation gap) of the population is replaced by offspring. Using this principle in a GA and replacing only one solution at a time per iteration leads to the steady-state GA (Whitley & Kauth, 1988; Syswerda, 1989; Davis, 1991; Sarma & De Jong, 1997). In a steady-state GA a new solution is created by recombination or mutation and replaces the current worst solution in the population. In addition to incorporating elitism in this strategy, this also allows a more efficient search because the current best solution is always kept in the population (Reeves & Rowe, 2003). The new offspring can immediately be used for the next search steps and there is no need to wait for completing a whole generation (Levine, 1996). However, in steady-state GAs the chance is higher that the population converges too early and that the landscape is not sufficiently explored, especially when dealing with small population sizes (De Jong & Sarma, 1993; Sarma & De Jong, 1997). Thus in practice, the population sizes of steady-state GAs are often very high.

3.5 Summary

In this section, some principles of metaheuristics were presented. Metaheurstics are general-purpose improvement heuristics that could be used to solve NP-hard problems to which exact optimization techniques can hardly be applied. Some problem-specific adaptions of the metaheuristic are necessary to achieve an efficient search process. This problem customization focuses on four basic design elements that every metaheuristic incorporates: the combination of representation and search operators, the fitness function, the initialization, and the search strategy. With respect to the search strategy, in general the different variants of metaheuristics can be classified into two groups: techniques based on local search and techniques using recombination-based search operators. For each search concept, one metaheuristic was specified as a representative example: threshold accepting as a local search routine and genetic algorithms as a recombination-based search. The choice of the search concept depends on the problem and its structure, however, real-world problems often have properties applicable to both search concepts, local and recombination-based search. Thus, the researcher either has to do an extensive study to make a final decision for one search strategy, or a metaheuristic using both types of search operators could be applied. Although presented as an example for recombination-based search, genetic algorithms also use a local-search operator in the background, and, thus, represent a promising technique for those real-world problems demanding both search concepts.

Chapter 4
Integrated Airline Scheduling

Abstract. Two approaches for integrated airline scheduling were presented and evaluated. They integrate the subproblems network design, frequency assignment, flight scheduling, fleet assignment, and aircraft routing. Furthermore, a schedule evaluation procedure was developed and calibrated that is required by both airline scheduling approaches. Both planning approaches are able to represent airline operations and practical requirements on a higher level of detail compared to many solution models presented so far. There are fewer simplifying assumptions or restrictions to certain planning scenarios. Their only requirement is to receive a quality measure for each schedule processed. The first airline scheduling approach follows the traditional sequential planning paradigm. This stepwise approach is realized in an iterative procedure consisting of solution models from literature. In contrast, the second planning approach represents a truly simultaneous model. In a self-adaptive metaheuristic, each processed solution represents a complete airline schedule, thus including all former subproblems implicitly. A comparison in which both approaches are applied to the same scenarios confirmed the postulated higher performance of a simultaneous optimization since the simultaneous approach outperformed the sequential approach with regard to the operating profit of the obtained schedules and the required computational effort. The capability of the simultaneous planning approach is further investigated by its application to scenarios that were modified implying a certain structure of the optimal solutions. For all experiments, the resulting schedules are in accordance with theoretical expectations.

4.1 Introduction

4.1.1 Motivation

The objective of airline scheduling is to create an airline schedule that is optimal with regard to a given objective value, usually operating profit. Because of the large number of variables and restrictions and their interdependencies that have to be

T. Grosche: Computational Intel. in Integrated Airline Scheduling, SCI 173, pp. 59–171.
springerlink.com

taken into account to solve this problem optimally, using exact OR techniques is computationally intractable. Instead, the airline scheduling problem is solved in a sequential planning process: the overall problem is decomposed into less complex subproblems that are solved in a sequential order.

Since its beginnings in the 1950s, numerous OR models for airline scheduling subproblems have been developed. With advances in optimization theory and computational power, improved solution approaches have been developed. Solution quality has increased and more realistic problems could be solved (in terms of level of detail and problem size). Additionally, researchers have begun to integrate subproblems into single optimization models to capture interdependencies that cannot be considered in a sequential solution approach.

However, there is still room for further improvements. Better solutions can be achieved if boundaries between subproblems of the airline scheduling problem are further relaxed. To obtain meaningful results, approaches must represent airline operations in higher levels of detail and include more practical requirements. Furthermore, optimization models must be capable of solving realistic-sized problem instances in a flexible way and not be restricted to specific problems or solution structures.

This study focuses on these challenges. Two solution approaches for integrated airline scheduling are developed, each combining the complete flight schedule generation and aircraft scheduling steps (see Fig. 2.2) into one model.[1] Both models do not have any specific requirements regarding the problem structure, thus, they can be used for any possible planning scenario and allow a flexible scheduling based on the given setup. In addition, both approaches are capable of representing airline operations on a high level of detail, allowing the inclusion of many practical requirements. Because even selected subproblems of the airline scheduling problem are NP-hard, the overall problem is NP-hard and cannot be solved in a single model using exact optimization techniques. Therefore, the two approaches represent iterative procedures incorporating heuristic elements. The first approach follows the traditional planning paradigm: in an iterative procedure, in each iteration subproblems are solved in a sequence. In contrast, the second model represents a simultaneous planning approach: complete airline schedules are optimized as a whole using metaheuristic optimization. The objective of both models is to maximize the operating profit of a daily airline schedule. Because both approaches use the same input data, a fair comparison with regard to the obtained solution quality and the computational effort is possible.

One important element of each planning approach is the quantification of the operating profit of a given airline schedule. Schedule evaluation itself represents a very complex problem, and many different (commercial) applications for schedule evaluation exist (for example Sabre® Airline Profitability Model, or United Airlines' Profitability Forecasting Model (Lohatepanont & Barnhart, 2004)). Because these tools use very sophisticated methods and take many parameters into account, their computation times vary from minutes to many hours. Using these models in each

---

[1] In Sect. 4.6.4 the possible integration of crew considerations is described.

airline scheduling approach would lead to excessive overall computation times. In addition, their acquisition costs, the lack of parameters and data required by those tools, and the general difficulty in obtaining detailed information due to their proprietary nature (Lohatepanont & Barnhart, 2004) make it impossible to use them for this study. Thus, an additional focus of this study is the development and calibration of a schedule evaluation procedure that is used by both the integrated airline scheduling approaches.

4.1.2 Structure

This introduction ends with an overview of the data necessary and used in all experiments. Then, four main sections follow. First, the schedule evaluation procedure for the airline schedule optimization methods is presented in Sect. 4.2. Then, sections 4.3 and 4.4 describe each planning approach in detail including the calibration of their parameters and an analysis. A comparison of both models is presented in Sect. 4.5. This section also applies the integrated scheduling approach to systematically modified scenarios to verify the obtained solutions and to demonstrate the capability of the planning approach.

4.1.3 Data

The objective of the airline scheduling approaches of this study is not only to integrate the subproblems of airline scheduling, but also to allow a flexible and general planning without limitations to certain scenarios (for example only (one) hub-and-spoke networks, one fleet type, etc.) on a detailed level. This is accomplished by taking practical requirements into consideration and using real-world data for calibration and testing. This section focuses on the latter requirement and presents an overview of the data used in this study. If reasonable and necessary, precise data values are included in the appendix. Without losing generality, the focus of this study is on the European airline system, thus, the data is limited to European figures.

Since airline scheduling includes a major portion of the overall operation of an airline, a significant amount of data is necessary. However, in general, data regarding airline operations represents highly classified information of airlines that is usually not accessible from outside, at least not to the required extent. Thus, in this study, some data represents real figures from airlines, some is obtained from industrial organizations and offices, and some data has had to be estimated or aggregated based on the fragmented information that was available. If all the necessary information were directly available, it could be used immediately.

In general, the information required to construct an airline schedule can be divided in general, demand, and supply data (see Fig. 4.1).

General data includes fundamentals that are unchangeable, for example the location of airports, operating characteristics of aircraft etc. Demand data focuses on the airline's market. This information cannot be influenced by an individual airline and includes passenger behavior and demand, competing airlines, etc. Supply data

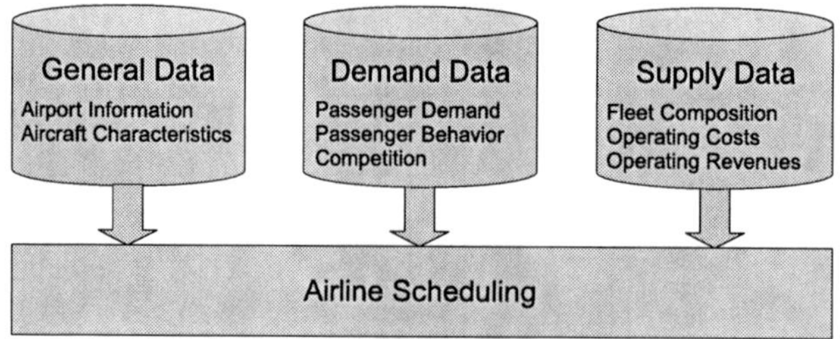

Fig. 4.1 Required data for airline scheduling

depends on the airline and its planning scenario, including for example the number of aircraft and the fleet composition, some operational characteristics and economic data like costs and revenues.

Before presenting the three different types of data in detail, two valuable sources of information need to be introduced that were used in this study: data from Official Airline Guide (OAG) and Market Information Data Tapes (MIDT). OAG is a global travel and transport information company that provides database administration in the area of aviation, travel and hotels (OAG, 2007). Its airline schedules database holds future and historical flight details for 1,000 airlines and more than 3,500 airports. For this study, historical schedules from 2004 (January to August) were available. After filtering and aggregating, this data set contains 3,274,756 different flight records, each with the following information:

- airline and flight number,
- origin and destination airport (city),
- aircraft and category (for example propeller, turbo-prop, jet),
- departure and arrival time,
- validity period,
- distance and elapsed time,
- number of seats (total, first, business, economy).

MIDT contain passenger bookings made via all the major global distribution systems (GDS). In this study, MIDT data was given for seven months in 2004 (January-April, June-August) for airline travels between Germany and European countries with non-stop or one-stop service. This data set consists of 1,365,497 records including a total of 7,808,041 bookings. The following information is part of each record of MIDT:

- origin and destination,
- connection airport (if applicable),
- airline and flight number of the first flight leg,
- airline and flight number of the second flight leg (if applicable),
- departure and arrival date and time of the first flight leg,
- departure and arrival date and time of the second flight leg (if applicable),
- booking class of the first flight leg,

- booking class of the second flight leg (if applicable),
- number of bookings.

4.1.3.1 General Data

The general information focuses on the *environment* the airline is in. This data is fixed and the same for every airline, no airline is able to influence or change any data. It is composed of characteristics of airports and aircraft.

Airport-Related Data. This study focuses on the European airline travel system. From the OAG flight schedules, 320 different airports were identified as having regularly scheduled airline services. Some of these airports are located at or near the same city (*multi-airport cities*), thus, the number of cities is only 307. One of the most important attributes of an airport is its location. The location is expressed as latitude and longitude coordinates.[2] The distance between any two airports is calculated as great circle distance based on the coordinates of the airports. A great circle divides a sphere into two hemispheres, thus, it has the same circumference as the sphere. The shortest path between any two positions on the surface of the sphere is part of a great circle. Given the coordinates *lat* and *lon* of two cities i and j, their great circle distance can be calculated using the following formula with $\Delta\sigma_{ij}$ as the angular difference:

$$\Delta\sigma_{ij} = 2\arcsin\cdot\sqrt{\sin^2\frac{lat_j - lat_i}{2} + \cos lat_i \cos lat_j \sin^2\frac{\Delta lon}{2}}. \qquad (4.1)$$

Besides their location, the second most important attributes of the airports for airline scheduling are their operating restrictions. For example, the length of the runway or layout of the apron might exclude certain aircraft types from service at that airport. For this study, the length of the longest runway is used as a potential restriction, appropriate data was determined for every airport using the ICAO World Airfield Catalogue (Woodside, 2000). Besides these operational limitations, many airports apply operating curfews or movement restrictions to comply with noise abatement procedures. The design of these curfew restrictions is manifold and many different procedures exist depending on the aircraft type and its noise certification. At most airports with curfew restrictions there is one period of time (usually at night hours) when the airport effectively is closed. In this study, only one closing period per airport is assumed, regardless of the aircraft type. Using multiple sources (for example Hochfeld et al. (2003), ACI (2004), EC (2005)), the operating hours of the 320 airports were investigated. Airports with no information available were assumed to have no curfew restrictions. Average values were calculated for inconsistent information.

Aircraft-Related Data. Fleet types differ in their characteristics that have to be taken into account when constructing an airline schedule (like seat capacity, fuel

---

[2] For example, Frankfurt Airport is located at 50°01'35.12"N 008°32'35.25"E.

consumption, cruising speed, range, minimum turn times etc.). Given the flight schedules from OAG, a total of 189 different aircraft types were identified including aircraft types of the same fleet family (like for example the Airbus A318/319/320/321 family). The total capacity of each aircraft type can be easily obtained using the information from the OAG flight schedules. Capacities were averaged for inconsistent information. A second characteristic of every aircraft type necessary in airline scheduling are operational limitations. In this study, these limitations include the range and the required landing distance. Without going into detail, many different sources were used to determine the range and landing distance for each fleet type, mainly online sources and internal reports. Because both values depend on many factors (for example weather, weight of the aircraft, runway surface and slope, altitude etc.), averages are used in this study. Appropriate data can be obtained for 38 different aircraft types, thus, limiting the usable fleets for airline scheduling to this number. Another important element for airline scheduling are block and turn times. These values depend on the fleet type and the flight the aircraft is assigned to. Given the OAG flight schedules, block times for different airport-pairs and fleet types can be easily obtained. As not every combination of airport-pair and fleet type were available, missing values were obtained by a regression model for each fleet type calibrated with existing information and using the distance of a city pair as independent variable. Turn times were given for certain aircraft types from various sources and can be used directly. Missing values were calculated using a regression model with the aircraft type's capacity as independent variable.[3]

The aircraft-related information is presented in Sect. A in the appendix.

4.1.3.2 Demand Data

In general, demand information includes airline demand in absolute passenger numbers, the passengers' preferences and travel behavior based on the offered flight schedules, and the competition between the airlines. The MIDT data represents a valuable source of demand information. However, because this data is limited to specific flights that took place in the past depending on the historical schedules of the airlines, it cannot be used directly for non-restricted planning scenarios in which different flights might take place. Instead, implicit information within the MIDT data is used to produce estimates of demand figures. The absolute passenger demand for airline travel is estimated using a gravity model (see Sect. 4.2.2), passenger preferences and behavior is reproduced using the passenger demand models in Sect. 4.2.4. OAG flight schedules provide information of competing flights. Because the focus of the integrated scheduling approaches of this study is on the daily airline scheduling problem, (connecting) flights are assumed to compete if they take

---

[3] The classification of turn times as general data should not hide the fact that airlines do have an influence on turn times by modifying ground operations. In addition, in general turn times also depend on the previous flight (for example, a long-haul flight requires longer refueling than a short-haul flight). However, a large portion of the required turn time is determined just by the fleet type. In addition, because only little information was available and the airline-specific impact could not be modeled in this study, turn times are assumed to be given as fixed data.

place on the same day in the same market. As competing flights, real flights from the OAG schedules are used; on average, there are about 14,500 flights per day in Europe that need to be considered as competing flights.

4.1.3.3 Supply Data

Supply information depends on the airline and its planning scenario. For example, the airline has a given number of aircraft of different fleets. Since in this study airline schedule optimization is not focused on a specific airline but might be applied to any planning scenario, there is no limitation or orientation to specific supply data. For each planning scenario, all 38 different fleet types and all 320 airports are available. The airports represent departure or arrival candidates for an airline's flights. In each planning scenario the set of airports can be reduced to those airports that are allowed as arrival or departure airports for the airline scheduling process.[4] From this set of airports, at least one airport per fleet type has to be selected as a maintenance station at which aircraft undergo maintenance on a regular basis. Since no information of maintenance facilities was availabe, airports are randomly selected as maintenance stations for each fleet type in order to conduct the planning scenarios in this research. It is assumed that one fleet type's maintenance station is capable of conducting maintenance for all fleet types of the airline with less seat capacity.

One important type of supply information is economic data. If the number of passengers on each flight is known, the revenue has to be estimated. The revenue is calculated by multiplying the number of passengers in each market with the average yield per passenger in this market. In reality, the yield differs between markets and airlines. Unfortunately, detailed information is not available to the public. In addition, airline fares are difficult to reliably forecast or aggregate, because they mainly depend on route competition and on different restrictions of the fare classes (Jorge-Calderón, 1997, Lee, 2003). AEA (2005) published an average amount of passenger revenue received per revenue passenger kilometer (RPK) of 17.0 US cents for airline travel in Europe. In this study, this value was multiplied by the distance between the cities of each market to produce yields per market. In addition to revenues, operating costs need to be calculated. They include those costs directly related to the airline schedule. Although cost assignment itself represents a complex topic and many different cost classifications exist, in this study a rather simple but practical and reasonable approach to cost assignment is used. Operating costs depend on many different factors but can be summarized to costs dependent on the aircraft type, flight routing, and the number of passengers. Costs depending on the number of passengers are excluded in this study because the yield per passenger already incorporates these costs. The remaining costs depend on the combination of aircraft type and a flight's airports (for example landing fees) and the combination of aircraft type and flight time (for example fuel, proportional maintenance costs etc.). Because there was no information of landing fees etc. available, these costs have

---

[4] In Sect. B in the appendix five different planning scenarios are presented that are used for the experiments of the airline scheduling approaches.

to be neglected. The other type of costs – block costs – are estimated using data from 2004 published by the Bureau of Transportation Statistics (BTS) of the US Department of Transportation. This department collects and publishes financial reports (BTS form 41) with financial information on large certified U.S. air carriers (BTS table P-51) and large and medium regional air carriers (BTS table P-52). This data includes detailed data of various cost categories for several US airlines on a flight, route, or aircraft level (BTS, 2006). Any invalid entries or obvious inconsistencies were removed manually. The overall operating expenses are given in most cases on a fleet type level for the 38 different fleet types. Dividing these costs by the total block hours reported for each type leads to block hour costs for each fleet type. Using the block hour costs and the block time for every pair of airports and fleet type, the total block costs can be calculated.

The block hour costs are presented in Sect. A in the appendix.

4.2 Schedule Evaluation

4.2.1 Overview

In airline scheduling, it is of utmost importance to assess schedule drafts with regard to their operating profit. While it is easy to use observed passenger numbers or profit shares of existing flights to decide about their quality, this information is of little use when planning new flights that are not part of past schedules for which passenger numbers are known. Thus, when optimizing an airline schedule, an evaluation model is necessary to assess the quality of a proposed schedule.

Many different passenger estimation tools exist that airlines use to estimate passenger numbers. These tools are provided by commercial suppliers (for example Sabre® Airline Profitability Model) or developed by airlines (for example United Airlines' Profitability Forecasting Model). Because these models try to reproduce passenger behavior on a detailed level, many different factors are taken into account that influence passenger demand. However, it remains questionable if a high number of parameters increases the level of accuracy of the predictions, because the calibration of each parameter inherits uncertainty. In addition, a large model requires high computation times, and the evaluation of a single airline schedule can last from minutes to hours. Because the airline scheduling models presented in this study primarily work iteratively with one or many schedule evaluations necessary in each iteration, existing evaluation models or rather commercial software could not be used, since this would lead to excessive computation times. In addition, due to their proprietary nature, little detail on these models is published, and their access is restricted to internal use. Thus, in this study, an airline schedule evaluation model has been developed that is used for the integrated airline optimization methods presented in sections 4.3 and 4.4.

Decisions on an airline schedule and its flights depend on the quality of the schedule and the individual contribution of every single flight to the overall quality. It is assumed that an airline wants to maximize its profit, thus, the quality of an airline schedule can be determined by its contribution to this overall goal. Although the

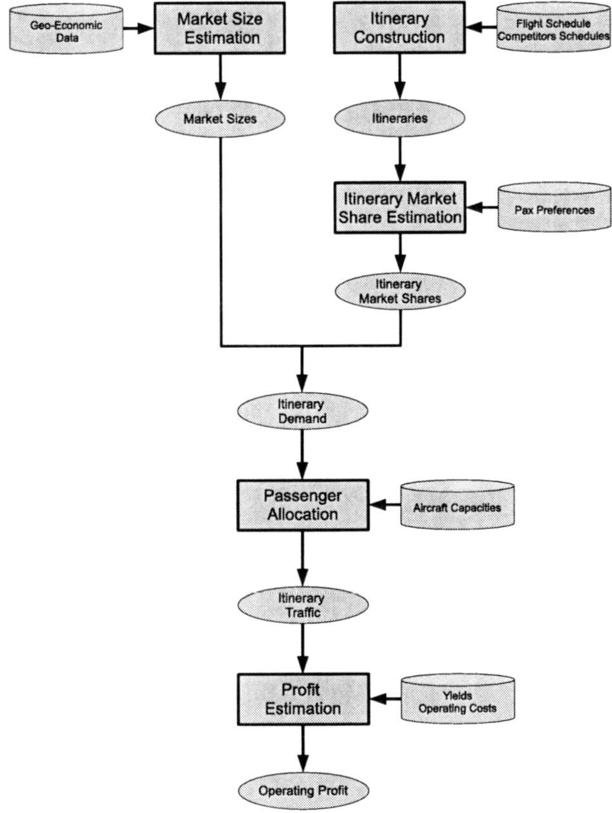

Fig. 4.2 Schedule evaluation process

profit depends on many different properties of an airline schedule, usually the operating profit is used to determine a schedule's quality. The operating profit is defined as the profit directly related to and dependent on the flights in the schedule, or, to be more specific, the yield of all passengers of all flights minus the costs for operating the flights.

Given an airline schedule, the operating profit is estimated using the following process:

1. *Market Size Estimation*: Estimation of the total number of airline passenger demand between any two airports.
2. *Itinerary Construction*: Selection and construction of itineraries (direct and connecting flights) that are offered to passengers.
3. *Itinerary Market Share Estimation*: Calculation of market shares or attractions of competing itineraries.
4. *Passenger Allocation*: Allocation of passengers to individual flights under consideration of capacity constraints.
5. *Profit Estimation*: Estimation of revenue and costs of all flights.

An overview of the process of airline schedule evaluation including the major required data is presented in Fig. 4.2. In the following, each step is presented in more detail.

4.2.2 Market Size Estimation

The objective of this step is to estimate airline passenger market sizes. A market is characterized by a city-pair and time interval (for example one day). Then, the market size denotes the total number of passengers (passenger volume) that intend to travel by air in this market. Because there is no data for market sizes given for this study, they have to be estimated using a forecasting model.

A variety of different forecasting techniques for market size estimation exists, and no technique can actually guarantee the accuracy of its predictions (Doganis, 2004). Even similar methods may produce widely diverging forecasts. Therefore, in practice an airline usually makes use of many forecasting models to increase the level of accuracy and trust, or simply uses numbers of past passenger flows as travel demand for airline scheduling. However, if this data is unreliable, not available, changes in demand structures exist, or new markets are evaluated, airlines need to estimate market sizes.

In this study, market sizes are estimated using a gravity model. Gravity models were the earliest causal models developed for human spatial interaction and traffic forecasting (Doganis, 2004). The use of a gravity model in this study is motivated by the fact that in gravity models it is assumed that there is a specific functional relationship between travel demand (as dependent variable) and the characteristics of the market (as the attracting and deterring independent variables). Thus, once calibrated, a gravity model can be used to estimate passenger flows for every market with its characteristics known. If the set of independent variables is carefully selected and the model properly calibrated, a gravity model can be used to estimate passenger volumes independently of the characteristics of present or past flights in the markets. This is a necessary prerequisite for passenger estimation the airline scheduling can rely on, because the flight characteristics are not given until the schedule is constructed. The gravity model used in this study represents a reduced variant of the gravity model developed by Grosche et al. (2007) which primarily uses a set of independent variables that do not depend on characteristics of existing or historic flight services.

In the following two sections, an overview and classification of forecasting techniques and drivers of passenger demand are given that can be used for passenger volume estimation. Then, in Sect. 4.2.2.3 gravity models are introduced study is presented in Sect. 4.2.2.4.

4.2.2.1 Forecasting Techniques

In general, there are two groups of forecasting techniques: qualitative and quantitative techniques (Doganis, 2004). Fig. 4.3 presents a classification of forecasting techniques.

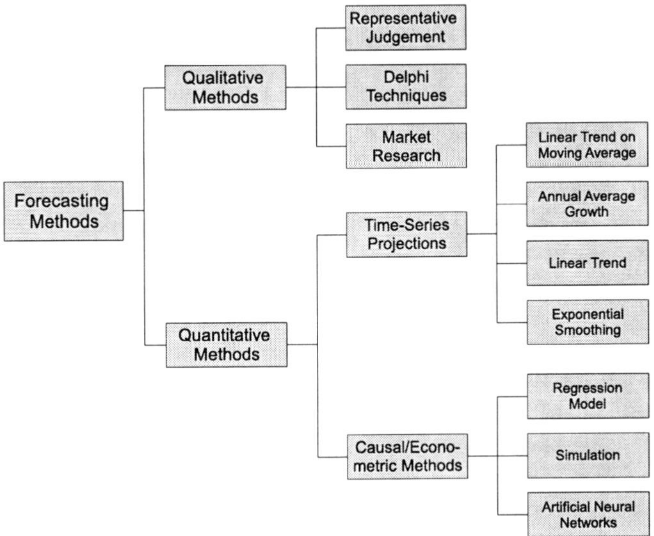

Fig. 4.3 Classification of forecasting techniques

Qualitative Techniques. Qualitative techniques include rough estimations and literal annotations of expected trends. Important techniques are executive judgment, market research, delphi techniques, and historic conclusion by analogy. One major advantage of qualitative techniques is their applicability if historical data is not available. On the other hand, techniques like market research, for example, might be a time-consuming and expensive task, and executive judgment relies only on the assessment of a few experts.

Quantitative Techniques. Quantitative models use mathematical models or relationships between independent and dependent variables. Formulating mathematical models allows independent estimations of future developments. Depending on the type of independent variables, two groups of techniques within quantitative models can be identified: causal/econometric methods and time-series projections.

Time-Series Modeling. In time-series or trend projections, a mathematical model is constructed with demand as a dependent and time as the only independent variable. As time progresses, airline demand will progress. It is assumed that the demand for air travel, the factors influencing demand, and their relationship remain stable and allow reliable forecasts. However, the rapidly changing business environment of airlines might violate this assumption. In addition, these kinds of models require accurate and sufficient historical data for model calibration for each route under investigation. Thus, on routes that are under evaluation for a new airline service, these models cannot be used because historical data is not available. The most important time-series projections include methods of annual average growth, exponential smoothing, linear trends, and linear trend on moving averages (Weatherford et al., 2003, Doganis, 2004).

Causal Modeling. Within causal methods, a functional relationship between air travel demand and selected economic or social supply variables is constructed. These models assume that airline travel demand can be derived from other factors and supports the realization of other targets like doing business or vacation trips (O'Connor, 1982). If one supply variable changes, air travel demand will also change. In econometric models, this relationship is measured, and by predicting changes in any one of the independent variables the impact on air travel demand can be forecasted (Doganis, 2004). In contrast to other forecasting techniques, individual independent variables and their influence on the dependent variable can be analyzed. Thus, causal models can help to evaluate different strategies and to investigate passenger behavior on a more detailed level. There are different models on how to construct the relationship between air travel demand and its influencing factors. A commonly used model is a regression model, since it is easy to construct and efficient calibration techniques are established. Due to their mathematical structure, gravity models can be considered as one variant of regression models. Causal methods also include simulation techniques and artificial neural networks. However, because the causal relationships cannot be extracted from these models, an explicit interpretation and analysis of the (functional) relationship of individual variables is difficult (see for example Weatherford et al. (2003)).

4.2.2.2 Drivers of Air Travel Demand

The first step in constructing a causal model is the identification of its variables. The travel demand as the dependent variable is usually measured as passenger volume describing the total number of passenger trips on city-pair routes during a given period of time (Kanafani, 1983). Because the total demand includes many individuals with different behavior characteristics, a higher level of accuracy can be achieved by forecasts for homogeneous groups of individuals. A common segmentation is by purpose of travel (business or leisure travel), since the demand for travel is usually derived from the demand for activities at the destination (Bouamrene & Flavell, 1980, Kanafani, 1983). The factors that influence the demand of both segments are assumed to be different in type and impact.

In general, factors that influence passenger volumes are included as independent variables and can be categorized into two groups: geo-economic and service-related factors (O'Connor, 1982, Kanafani, 1983, Rengaraju & Thamizh Arasan, 1992, Jorge-Calderón, 1997). Geo-economic factors are factors that describe the economic activity and geographical characteristics of the airports and cities of the route under investigation. Service-related factors include characteristics of the air transport system between two cities. In contrast to geo-economic factors, service-related factors are under the control of the airlines.

Geo-Economic Factors. Geo-economic factors include activity-related factors that describe the (economic) activity of a city or between a city-pair, and their geographical characteristics.

The most commonly used activity-related factors are income and population of the metropolitan area served by an airport because these are useful approximations of activity factors (Kanafani, 1983). An even more aggregated measure of the economic activity and income levels is the past total passenger volume at each airport (Doganis, 1966). More detailed activity-related variables that have been used earlier include income distribution, percentage of university degree holders, number of full-time employees, type of city, employment composition, structure of the production sector of one region, or economic, political and cultural relationships between two countries (Kanafani, 1983, Russon & Riley, 1993, Jorge-Calderón, 1997).

Geographical factors are characteristics of the location or geographical properties of a city or city-pair. For example, an important factor for inter-city air travel demand is the distance between cities. Two opposite effects of the distance of a city-pair on its demand can be identified. First, with increasing distance less social and commercial interaction can be observed. Second, long distances increase the competitiveness of air transportation over other transportation modes regarding travel time, especially when there is no overland connection is available (O'Connor, 1982, Kanafani, 1983, Russon & Riley, 1993, Jorge-Calderón, 1997). The competition of airports in close proximity of one of the airports of the route under investigation is also considered as a geographical factor on demand (Russon & Riley, 1993). For example, an airport with a better schedule in terms of frequency or destinations is likely to offer more convenient departures and, thus, might attract more passengers than an airport in close distance (Fotheringham & Webber, 1980, Fotheringham, 1983, Ubøe, 2004).

Service-Related factors. In general, the service of air travel is determined by its quality and its price (Jorge-Calderón, 1997).

Existing studies show that there are many different factors that influence airline service quality (Ippolito, 1981, Ghobrial & Kanafani, 1995b, Wojahn, 2002, Gardner Jr., 2004, Gursoy et al., 2003, Park et al., 2004). An important factor is the total travel time between two city pairs which is determined by the desired departure time of a passenger and the actual arrival time. The overall travel time also depends on the frequency of flights offered in a market (Kanafani, 1983, Jorge-Calderón, 1997, Proussaloglou & Koppelman, 1999). With increasing frequency passengers are able to select a flight that departs closer to their preferred departure time minimizing their total travel time. The average load factor also influences overall travel time as it indicates the probability of free seats at the preferred departure time. As flight delays increase the travel time of passengers, the overall on-time performance of an airline is another factor. An airline's reputation is also important for the service quality, as well as the market presence, customer loyalty programs, and the aircraft equipment (Kanafani, 1983, Jorge-Calderón, 1997, Proussaloglou & Koppelman, 1999).

The relationship between price and demand has been studied in various publications. In general, the demand for air travel decreases with increasing fares. Especially on short-haul routes airlines face competition by other transportation modes that gain a relative advantage with increasing air fares (Russon & Riley, 1993, Jorge-Calderón, 1997). The results of a survey of German passengers showed that 52% of all passengers would not have traveled at all if no low-priced flights (offered by a low-cost airline) had been available (Tacke & Schleusener, 2003). On the other hand, some researchers reject the consideration of air fares in travel demand forecasting models. Their reasons are manifold. In many cases, the air fare is highly correlated with the distance or travel time and, thus, should not be considered as an independent factor (Kanafani, 1983, Rengaraju & Thamizh Arasan, 1992). Another reason is that air fare is assumed to be an exogenous factor (Jorge-Calderón, 1997); an airline has only limited control over the prices it charges because it must meet the same prices as potential competitors (O'Connor, 1982). In addition, it is difficult for airlines to reliably forecast fares because the important determinants like oil prices are highly volatile and hard to predict (Doganis, 2004). Finally, the use of average fares in forecasting models is problematic because air fares mainly depend on route density/competition and on different restrictions of the fare classes (Wells, 1998, Lee, 2003).[5] For example, Jorge-Calderón (1997) showed that air travel demand is price inelastic with respect to the unrestricted economy fare, and moderately discounted restricted fares do not significantly generate additional air traffic. Even though there exists inter-modal competition in short-haul routes, airline service is usually used by passengers that are time-sensitive and price-insensitive.

4.2.2.3 Gravity Models for Air Traffic Forecasting

Introduction. Gravity models are inspired by the gravitational law of physics (Newton, 1687). The gravitational law states that the gravity between two objects is directly proportional to their masses and inversely proportional to their squared distances.

A simple formulation of a gravity model for human spatial interaction used for the prediction of traffic flows between two cities i and j can be formulated as

$$V_{ij} = k \frac{(A_i A_j)^{\alpha}}{d_{ij}^{\gamma}}, \tag{4.2}$$

where V_{ij} is the passenger volume between i and j ($i \neq j$), A_i and A_j are attraction factors of i and j, d_{ij} is the distance between both cities (or any other impeding factor), and k is a constant. γ is a parameter that controls the influence of the distance on travel demand and α controls the influence of the attraction factors.

Several extensions to this simple formulation exist. If V_{ij} is measured by passenger volume originating from i and ending in j (instead of the total two-way traffic), separate variables can be included representing travel production factors (push factor) P_i of the originating city and travel attraction factors (pull factor) A_j of the destination city and individual parameters α and β controlling their influence. This

---

[5] In the USA, more than 30,000 fare changes per day are observable (McGill & Van Ryzin, 1999). Thus, historic airline fares are of limited use at the time of airline scheduling requiring market size estimation, since this planning task is conducted well ahead of the day of operation. The closer the departure, the more important are airline fares. In fact, airline fares represent the short-term instrument for capacity control, while the airline schedule serves as a medium or long-term instrument.

distinction is sometimes made only by allowing the variables to have different parameter values for the origin and destination city while using the same variables for both (Kanafani, 1983). In some approaches, α and β are city-specific (denoted as α_i and β_j). A gravity model can be constrained by production or attraction. In attraction (production)-constrained models, the total travel demand attracted to one city (produced at one city) has to be equal to the observed arriving (departing) passengers at this city. A double-constrained gravity model (constrained in production and attraction) can be formulated as

$$V_{ij} = \frac{P_i^{\alpha_i} A_j^{\beta_j}}{d_{ij}^{\gamma}}. \tag{4.3}$$

The production and attraction of each city is constrained:

$$\sum_{j}^{n} V_{ij} = P_i \tag{4.4}$$

and

$$\sum_{i}^{n} V_{ij} = A_j. \tag{4.5}$$

In this kind of model, it is assumed that the total demand or flow leaving a city is known and this knowledge is used within model formulation and calibration. However, a constrained model cannot be used for demand prediction in new markets because calibrating the constraint model is not possible if no historical data is available.

Calibration. After defining the basic model formulation, the parameters of the model have to be calibrated. The objective of calibration is to find model parameters that lead to an accurate prediction of the expected travel demand (the gap between predicted travel demand and observed travel demand should be small). The calibration can be conducted by using either time-series or cross-sectional data (Kanafani, 1983, Rengaraju & Thamizh Arasan, 1992, Doganis, 2004). If time-series data is used, a demand model is calibrated for a particular city-pair using data for a number of different time periods. All variables in a time-series model are expressed as functions of time polynomials. This permits the construction of a demand function specific to each city-pair. In cross-sectional calibration, the same model is assumed to hold for a number of different city-pairs. The data for these city-pairs during one single time period is aggregated and used for parameter estimation (Moore & Soliman, 1981, Kanafani, 1983, Rengaraju & Thamizh Arasan, 1992).

In most cases, the calibration itself is conducted using the ordinary-least-squares method. Taking the logarithm of a gravity model formulation leads to a multiple linear regression expression to which standard techniques can be applied.

Previous Work. The variety of different independent factors allows the formulation of a large number of different gravity models. The following Table 4.1 presents some gravity models from literature that were tested cross-sectionally with real-world data and for which information on variables, number of observations (obs.), and coefficient of determination (R^2) as quality measure were published.

Table 4.1 Properties of selected gravity models from literature

Author (Year)	Factors	Obs.	R^2
Doganis (1966)	Observed passenger number at airports, distance	22	0.740^a
Brown and Watkins (1968)	Income, sales competition, average fare per mile, journey time per mile, number of stops, distance, phone calls, international passengers on domestic flights, competition index	300	0.870
Verleger (1972)	Income, price, phone calls, distance, flying time	441	0.720^b
Moore and Soliman (1981)	Population on city-level, income, economy fare	69	0.370
	Population of airport catchment regions, income, airport catchment, economy fare	58	0.810
Fotheringham (1983)	Attractiveness/population, traffic outflow of origin, distance	9900	0.730; 0.760
Rengaraju and Thamizh Arasan (1992)	Population, percentage of employees, university degree holders, big-city proximity factor, travel time ratio (travel time by rail divided by travel time by air), distance, frequency of service	40	0.952
Russon and Riley (1993)	Income, population, highway miles distance, number of jet/propeller nonstop/connection flights, driving time minus connection flight time, distance to competing airports, political state boundary	391	0.992
O'Kelly et al. (1995)	Nodal attraction, distance	294	0.850^c
Jorge-Calderón (1997)	Population, income, proximity of hub airport, hub airport, distance, existence of body of water between cities	339	0.371
	Additional variables: tourism destination, frequency, aircraft size, economy fare (not/moderately/highly discounted/restricted)	339	0.722
Shen (2004)	Nodal attraction, impedance	600	0.568^d
Doganis (2004)	Scheduled passenger traffic at airports, economy fare, frequency	47	0.941
Grosche et al. (2007)	Population, catchment, buying power index, gross domestic product, distance, average travel time	956	0.761
	Additional variables: number of competing airports, average distance to competing airports, number of competing airports weighted by their distance	1228	0.730

[a] This value is the *rank coefficient*. The city-pairs are ranked according to the actual and estimated passenger volumes and the correlation between the ranks yields the rank coefficient.

[b] The study is based on the model of Brown and Watkins (1968).

[c] The authors present different methods for a reverse calibration of the gravity model. The attraction of cities is estimated based on observed traffic flows. A multiple linear regression results in $R^2 = 0.850$, two different linear programming models result in $R^2 = 0.700$ and $R^2 = 0.810$.

[d] The focus is on an algebraic approach for reverse-fitting of the gravity model. Therefore, the nodal attraction is estimated endogenously from exogenous spatial interaction and impedance.

4.2.2.4 Gravity Model Development

The gravity model by Grosche et al. (2007) is used as a basis from which the model used in this study is derived. This model primarily uses geo-economic variables as input and cross-sectional data for calibration, allowing an application to new markets for which historical data is not available.[6] Statistical tests on the model show

---

[6] This set of variables is extracted from the data also available for this study.

satisfying results; in addition, the model was thoroughly validated with a formal test and by analyzing the stability of the coefficient of determination R^2 and the coefficients of the independent variables for different subsets of the total number of observations.

One reason for the good performance of this gravity model is its application to homogeneous data, because only routes between Germany and other European countries were available for calibration. This assumption is confirmed when applying the calibrated model to all city-pairs within Europe. Because some statistical offices and industrial organizations (for example Eurostat, AEA, ICAO etc.) publish traffic figures for some selected routes, their order with regard to passenger flows can be compared with the order of the estimated passenger demand figures. A comparison unveils an overestimation by the gravity model especially for long-haul routes. The reason for this might be Germany's central geographical position in Europe, leading to calibration data with markets representing typical medium-haul flights.

These observations support the basic requirement to apply gravity models to homogeneous markets (Kanafani, 1983). For this study in which all city-pairs in Europe are considered, the next step would be to build a set of gravity models and to calibrate them across the various markets within Europe, or to use European-wide data for the calibration of a more robust gravity model that then could be applied to all markets. In fact, the estimation of market sizes by airlines consists of the construction and calibration of many different models for individual regions or routes. Unfortunately, reliable data needed for such a calibration is not available for this study. Thus, two strategies remain to obtain market sizes for further use:

1. Usage of observed passenger flows as market size between those city-pairs for which this information is published.
2. Construction, calibration and application of a gravity model that has been reduced compared to the model of Grosche et al. (2007) to better reflect all markets in Europe.

The first strategy provides accurate data of realized passenger flows. However, these flows represent constrained passenger demand, because they result from existing airline services with their characteristics. For example, if there is no airline service on a market, the resulting passenger flow is zero even if demand exists. Or if capacities are small for a city-pair, the passenger flows probably underestimate the real unconstrained demand. Also because of the lack of available data, using this strategy results in a zero matrix for the demand between city-pairs with only a few cells filled. In contrast, a gravity model produces demand estimates with traffic between many city-pairs. Thus, it better reflects the (unconstrained) demand structure, although the individual passenger numbers estimated for the city-pairs will differ from the real values and the overall model fit might be poor. The reason for this is that the gravity model is applied to heterogeneous markets, although it was calibrated with homogeneous data sets. Nevertheless, this strategy is applied to produce market size estimates, because to assess different airline schedule construction techniques, the accuracy of passenger forecasts on selected markets is less important than considering more realistic demand structures across all markets. It has to be emphasized that this reduction and the related drawbacks result from the lack of information on the demand. If market sizes are available or could be obtained with any other estimation technique, this data could be used immediately to replace the estimates used here for schedule evaluation or to calibrate better fitting gravity models.

The gravity model used for market size estimation in this study is the basic model presented by Grosche et al. (2007) without the independent variables travel time and GDP. Airports of multi-airport cities were aggregated. The final model was manually selected by ordering the markets according to their estimated market sizes and comparing this order with the order of markets with real data available. It has the following form:

$$V_{ij} = e^{\varepsilon} P_{ij}{}^{\pi} C_{ij}^{\chi} B_{ij}^{\beta} D_{ij}^{\delta}, \qquad (4.6)$$

where V_{ij} is the total passenger volume between cities i and j, the exponents in Greek letters are used to model the impact of the input factors and are subject to the calibration process. The variables in capital letters are the independent factors influencing the travel volume. Table 4.2 lists the variables and their aggregate functional forms.

Table 4.2 Independent factors of the gravity model

Variable	Functional Form	Factor
P_{ij}	$P_i P_j$	Population
C_{ij}	$C_i C_j$	Catchment
B_{ij}	$B_i + B_j$	Buying power index
D_{ij}		Geographical distance

The following items briefly describe the independent variables used in the model.

- **Population:** The population of a city is determined based on various statistical offices of the involved countries. In all cases the latest figures were considered. The population refers only to the city of each airport, potential passengers from an airport's vicinity are included in the catchment data.
- **Catchment:** A catchment area of an airport covers the vicinity of an airport. Usually, the catchment area includes only those areas that are within a certain traveling distance to the airport. Consequently, the catchment area of an airport is defined as the region that is within 60 minutes driving time. The number of people living in this region are expected to use the airport for their travel purposes and are thus included in the catchment. The catchment data is derived from population data of the regions given on the NUTS3-level. NUTS (Nomenclature des unités territoriales statistiques) are classification levels of territorial units of about the same population size that provide the basis for regional statistics for the European Union (see Fig. 4.4 for an example). In this model, catchment data from the year 2003 is considered.
- **Buying power index:** The average buying power index is constructed on the basis of an airport's catchment area. Like the catchment, the buying power index

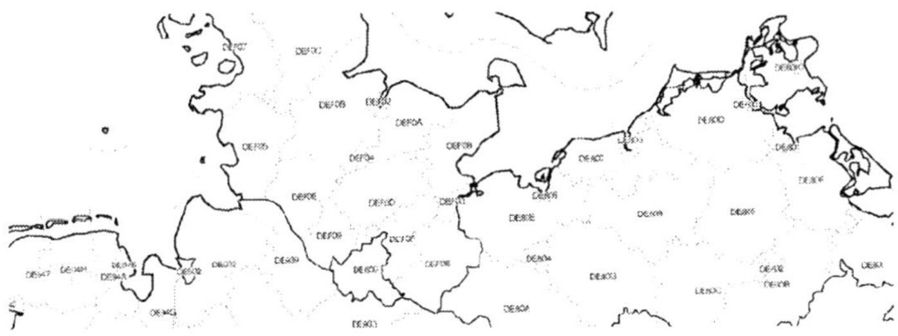

Fig. 4.4 NUTS3 on the northern coast of Germany

is given on the NUTS3-level with 100 as the European average. The index can
be interpreted as an indicator for the size of the travel budget of the population
within an airport's catchment. The data used for calibration is from 2003.

- **Geographical distance:** The distance between two airports is calculated as the
  great circle distance in kilometers between the airports' coordinates.

The calibration of the gravity model is conducted using the ordinary least square
method. Table 4.3 presents the coefficients of the resulting model with t-statistics
and standardized beta coefficients. The t-statistics indicate that the null hypothesis

Table 4.3 Calibration results of the gravity model

Coefficients	P_{ij}	C_{ij}	B_{ij}	D_{ij}
Values	0.357	0.203	1.722	-0.127
t-statistics	12.871	8.013	8.255	-2.047
Beta coefficients	0.350	0.229	0.222	-0.057

(the independent variables have no effect) can be rejected for each variable at the
5%-level. As discussed before, although the R^2 is rather poor ($R^2 = 0.283$), this
model is used to calculate market size, because it better reflects demand structures
across various markets in Europe.

4.2.3 Itinerary Construction

4.2.3.1 Overview

When constructing an airline schedule, the airline has to make the most efficient use
of its resources while best meeting the passengers' demand. The
compromise of these conflicting goals is the schedule presented to the potential
passengers containing travel itineraries. An itinerary is a travel alternative between
two cities, which could be either a direct flight or a sequence of connecting flights
(Coldren et al., 2003). Fig. 4.5 presents an excerpt from a CRS screen as an example
of four different itineraries presented to passengers for a journey between Hamburg
(HAM) and New York (JFK).

```
FLTNR            EFFECTIVITY          A/C A/C A/P  PT         A/P PT       S T
CAR    NUM       FROM   TO    OPSDAY  CAR TYP FROM D   STD MI  TO   A  STA MI T R CONF

UA     333       13MAR 13MAR 1234567      744 HAM  A   0800    JFK     1040      CY

LH     101       13MAR 13MAR 1234567      320 HAM  A   0800    FRA     0910      CY
LH     400       13MAR 13MAR 1234567      744 FRA  A   1000    JFK     1240      CY

LH     102       13MAR 13MAR 1234567      320 HAM  A   0800    CDG     0935      CY
AF     222       13MAR 13MAR 1234567      744 CDG  A   1030    JFK     1310      CY

LH     101       13MAR 13MAR 1234567      320 HAM  A   0800    FRA     0910      CY
AA     767       13MAR 13MAR 1234567      744 FRA  A   1040    JFK     1320      CY
```

Fig. 4.5 Itineraries between Hamburg (HAM) and New York (JFK)

It contains four travel itineraries departing at 08:00 from HAM: one direct flight and three connecting flights. By offering itineraries consisting of multiple flights, the airline is able to offer many more city-pair connections than with the given number of individual flights and, thus, can increase its passenger numbers. On one day, a city pair might be served by many different itineraries and a single flight might serve as a flight leg of many different itineraries. For example, in Fig. 4.5, the flight with flight number 101 is a segment of two itineraries.

Because each flight is a direct itinerary but might also serve as a segment of many different itineraries connecting different city pairs (markets), the estimation of passenger numbers cannot be conducted on a flight level. In addition, airline fares are given on a market level. Thus, for schedule evaluation, passenger flows have to be estimated on the itinerary level. Therefore, it is necessary to construct valid itineraries from the set of individual flights in the schedule. Because each flight of the schedule itself represents a valid itinerary, the focus is on the construction of connecting flights (or simply *connections*). In theory, the number of possible connections grows exponentially with an increasing number of flights, since all flights can be connected. Thus, the objective is to limit the number of connections to those itineraries that are reasonable and are likely to be chosen by passengers.

In practice, the itinerary construction process is conducted by CRS for passenger marketing (Coldren et al., 2003) and by airlines for direct sales and schedule evaluation. In these processes, many different algorithms or sets of rules are applied to remove infeasible, unrealistic or unreasonable itineraries. However, these rules and their parameters represent confidential information and there is no detailed information published that could be implemented for this study. However, there are some few comments about the general attributes of a connection to be considered as a valid itinerary (see for example Mathaisel (1997), Schmitz et al. (1998)). Based on this information, a set of rules for connection building was developed for this study. This set is presented in the following. To find all valid connections based on a given flight schedule, all rules are applied. As the rules incorporate some parameters, a calibration is necessary. A simple calibration was conducted using historic booking data from connecting flights; its results are presented in Sect. 4.2.3.3.

4.2.3.2 Connection Building Rules and Parameters

In the following, the set of rules used in this study for connection building is presented including parameters that have to be calibrated. Each paragraph focuses on one rule. To illustrate the functionality of each rule, the following Fig. 4.6 is introduced.

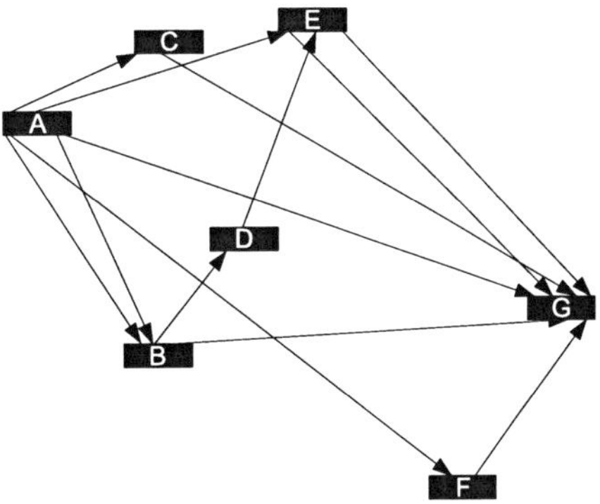

Fig. 4.6 Connection building example

In this network, the arcs represent single flights, the solid bars represent airports and their locations. For example, there is one flight scheduled between A and G, and airport D is located approximately halfway between A and G. The bars also represent a time-line (from left to right), allowing one to specify departure and arrival times of flights at each airport. For example, when looking at airport E, the first flight arriving departed from A, then a flight departs heading to G, a flight arrives from D, and the last flight departs towards G. In the following, the connection building rules are illustrated for connections between A and G.

Detour. In general, each connection itinerary represents a detour compared to the direct route between two cities. A maximum detour factor d_{max} indicates to what extent the geographical distance of a connecting flight $dist_{cnx}$ might exceed the direct distance $dist_{dir}$ between the origin and destination airport: $dist_{cnx} \leq d_{max} \cdot dist_{dir}$. This factor can be interpreted as an elliptical envelope around the direct route between two cities in which a connecting airport has to be located (Mathaisel, 1997). For example, in Fig. 4.7 routes via F between A and G exceed the maximum detour factor compared to the direct route between A and G.

In contrast, for example itineraries following the route via B or E are valid with regard to the detour factor.

Because it is based on geographical distance, the maximum detour factor reduces the number of possible connection itineraries independently of the flights in the schedule. Using the factor leads to a preselection of possible routes that valid itineraries have to follow. The separation of geographical routes and itineraries of the schedule allows an efficient implementation: the route selection needs to be performed only once with the complete set of available airports and is independent of the schedule that has to be evaluated in each iteration of the optimization approach.

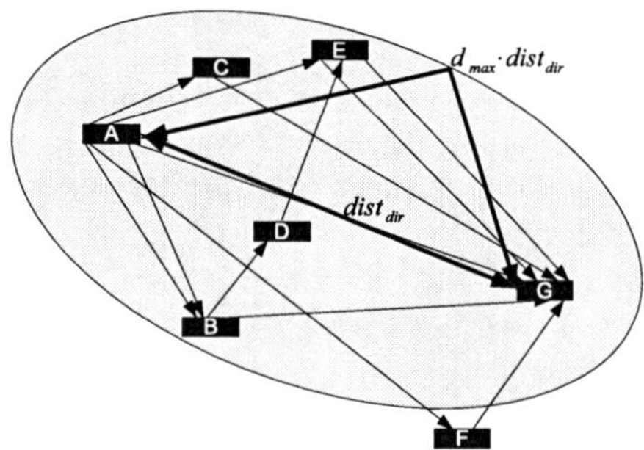

Fig. 4.7 Maximum detour factor d_{max}

Minimum Connection Time. To represent a feasible connection, the departure airport of a succeeding flight has to be the same as the arrival airport of its predecessor. In addition, the second flight must not depart before the first one has arrived and a minimum connection time t_{min}^{cnx} has elapsed. The minimum connection time is necessary for passengers to change the flights and to process their baggage between the two aircraft. Usually, the airport specifies this time.

Fig. 4.8 illustrates the application of this rule. For example, there is no connection flight A–D–G. In addition, although using the same airport, the first flight arriving at E cannot be connected with the first flight departing from E, because the connection time would exceed the minimum connection time t_{min}^{cnx}.

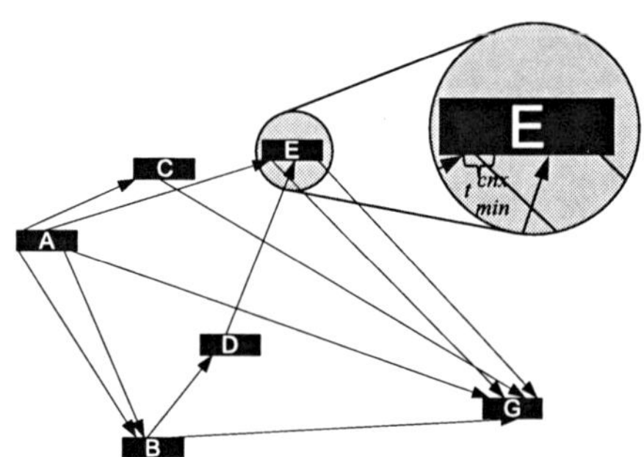

Fig. 4.8 Minimum connection time t_{min}^{cnx}

Number of Stops. In general, passengers want to minimize their travel time and to increase the convenience of the journey. Thus, if there are too many stops within a sequence of flights, this sequence is unlikely to be chosen by a passenger. A maximum number of stops s_{max} is defined that excludes connection flights with more intermediate stops than this number. If for example only one connecting airport is allowed ($s_{max} = 1$), in Fig. 4.9 there is no itinerary via D from A to G, since at least three intermediate stops would be required (via B, D, E).

Time Delay. In general, the shorter the travel time of an itinerary compared to competing itineraries, the more attractive it is to potential passengers. Thus, the

Fig. 4.9 Maximum number
of stops s_{max}

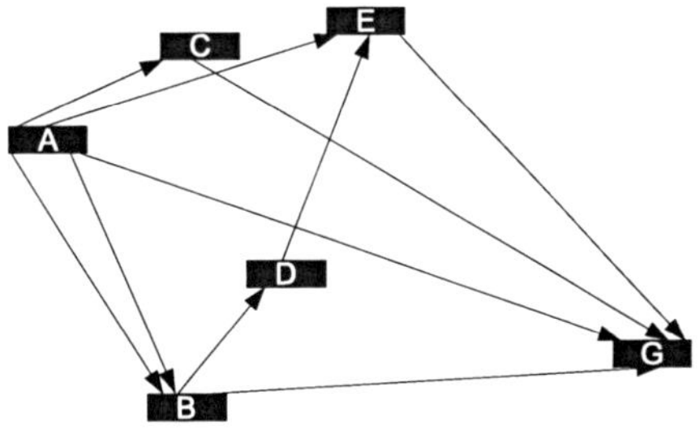

set of itineraries needs to be further reduced by removing itineraries with longer
travel times. A possible connection is excluded if its travel time t^{cnx} exceeds the
travel time of the shortest itinerary $t^{shortest}$ in the market by a certain factor t_{delay}:
$t^{cnx} \leq t_{delay} \cdot t^{shortest}$. It is assumed that the time delay is perceived differently depending on the type of the shortest itinerary (direct or connection), thus the factor
differentiates between both types. The time delay factor with regard to the shortest
connection itinerary is denoted as t^{cnx}_{delay}, the one with regard to the shortest direct
flight – if one exists – as t^{dir}_{delay}.

Fig. 4.10 presents the application of the described rule. There are two connections
via B from A to G. Depending on the time delay factor, the first flight between A
and B is likely to be removed from the connection building, since at least the second
connection itinerary promises a much faster connection.[7]

Fig. 4.10 Time delay factor
t_{delay}

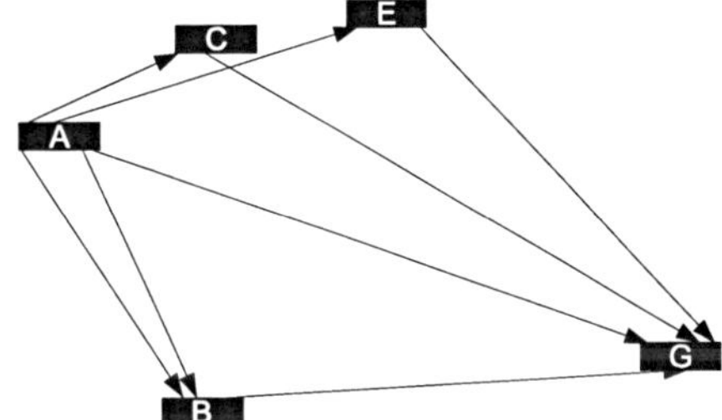

Maximum Connection Time. In contrast to the minimum connection time t^{cnx}_{min}, a
maximum connection time t^{cnx}_{max} specifies the time a passenger is willing to wait for
the connecting flight. A connection is only constructed, if the departure time of the
second flight leg departs before t^{cnx}_{max} has elapsed.

Fig. 4.11 illustrates the application of this rule. At airport E, the time between
the arrival from the flight arriving from A and the flight heading to G exceeds the
maximum connection time t^{cnx}_{max}.

---

[7] The factor must also take the connection flights via C and E into account. Furthermore, t^{dir}_{delay}
has to be considered, since there is a direct flight between A and G. However, for simplicity
reasons, these considerations are neglected in this example.

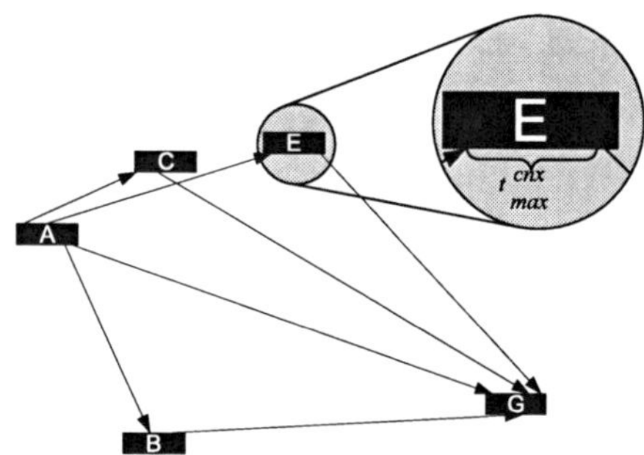

Fig. 4.11 Maximum connection time t^{cnx}_{max}

Interline Connections. Usually, airlines and CRS limit the number of connections of flights conducted by different airlines. Flights of the same airline are allowed to be connected without any restrictions (online connections). Flights of different airlines (interline connections) are usually only connected if the airlines are members of the same strategic alliance or have any other sort of interline or code-share agreement. With 210 different airlines from the OAG schedules, investigating all interline agreements would be beyond the scope of this work. Instead, in order to include interline connections, a parameter $n_{interline}$ is specified: if for any two airlines a number of interline connections exceeding this parameter could be observed in the given MIDT data, interline connections are allowed for these two airlines in future use.

4.2.3.3 Parameter Calibration

In the previous section, the parameters necessary for building connections were introduced. Each parameter needs to be calibrated to obtain a complete connection building procedure that reproduces the real travel behavior in the best way. This behavior can be observed using the given MIDT data, because each record represents a chosen itinerary and the number of passengers that have chosen this itinerary. Given all flights included in this data, the calibrated connection building sequence should result in a set of connections that were chosen by the passengers in the past. Because this number could easily be maximized by constructing all possible connections (leading also to a vast number of non-chosen connections and to excessive computation times), a second (conflicting) objective is to minimize the total number of connections constructed.

This calibration process represents an optimization problem. In this study, a simple metaheuristic is used to accomplish this task. To apply a metaheuristic to a given optimization problem, according to Rothlauf (2006a) the following four basic elements have to be adapted (see Sect. 3.2 on page 48):

1. **solution representation and variation operators**
   The seven parameters described in the previous section are the decision variables of the calibration process.[8] Thus, the genotype of a solution s consists of

---

[8] The maximum number of stops is set to $s_{max} = 1$ because only MIDT data with itineraries with a maximum of one stopover was available.

seven elements p each encoding one parameter as a continuous value. A local search operator randomly changes one value per solution step. The element p that is subject to the modification is selected randomly (following a uniform distribution). Then, the modified parameter p^* is calculated by multiplying p with a random value selected according to a normal distribution $N(1,0.5)$ with mean $\mu = 1$ and variance $\sigma^2 = 0.5$. This procedure leads to a higher probability for smaller changes. All parameters are protected against infeasible values (for example, detour factor $d_{max} < 1$). Since the metaheuristic is solely based on local search, recombination-based search operators were not implemented.

2. **fitness function**

   The fitness function has to integrate two conflicting objectives: maximization of the number of constructed connections that were chosen in the past and their passengers, and minimization of the total number of constructed connections. Let c_{os} denote the number of observed connections (index o) that were also selected (index s) by the connection builder, p_{os} the number of passengers traveling on these connections, c_{ot} the total number (index t) of observed connections, p_{ot} the total number of passengers, and c_{total} the total number of connections constructed. The term $\frac{c_{os}}{c_{ot}}$ expresses the number of connections constructed that were chosen in the past and has to be maximized. $\frac{p_{os}}{p_{ot}}$ is the corresponding number of passengers. Then, the average values of both terms are calculated as $\frac{\frac{c_{os}}{c_{ot}} + \frac{p_{os}}{p_{ot}}}{2}$. The objective to minimize the total number of constructed connections can be expressed by $\frac{c_{os}}{c_{total}}$. Using the parameters p_1 and p_2 to control the influence of each objective value on the overall fitness value, the fitness function is as follows:

$$f(s) = p_1 \cdot \frac{\frac{c_{os}}{c_{ot}} + \frac{p_{os}}{p_{ot}}}{2} + p_2 \cdot \frac{c_{os}}{c_{total}}. \tag{4.7}$$

   In this study, the two conflicting objectives are presumed to be equally important, thus the corresponding parameters are set to equal values ($p_1 = p_2 = 1$).

3. **initialization**

   As initial solution, the parameters are set to the following values:

$$d_{max} = 1.5,$$
$$t_{delay}^{dir} = 1.5,$$
$$t_{delay}^{cnx} = 1.5,$$

$$t_{max}^{cnx} = 120 \text{ minutes},$$
$$t_{min}^{cnx} = 45 \text{ minutes},$$
$$n_{interline} = 80.$$

4. **search strategy**

   A simple hill climbing technique is used as a search strategy. Thus, only local search steps are performed without accepting inferior solutions during the search process. If the algorithm does not improve for 500 iterations, it is terminated. The use of this rather simple optimization algorithm can be justified by preliminary tests, in which a threshold accepting algorithm did not result in better solution quality but much higher computation time. Algorithm 3 specifies the hill climbing technique used here.

Algorithm 3. Hill Climbing Algorithm for Parameter Calibration

1: create initial solution s with fitness value $f(s)$

2: iteration $i = 0$

3: **repeat**

4:    $i = i + 1$

5:    create new solution s^*

6:    calculate new fitness value $f(s_o^*)$

7:    **if** $f(s_o^*) > f(s)$ **then**

8:       $s = s^*$

9:       $i = 0$

10:    **end if**

11: **until** $i = 500$

Table 4.4 Connection building calibration results

Parameter	Value	σ
d_{max}	1.265	0.415
t_{delay}^{dir}	1.392	0.358
t_{delay}^{cnx}	1.741	1.158
t_{max}^{cnx}	113.099	44.767
t_{min}^{cnx}	55.067	24.747
$n_{interline}$	81.257	11.270
Fitness $f(s)$	1.414	0.365

Because the airline scheduling in this study is on a daily basis, the parameter calibration is conducted for each day separately. The following Table 4.4 presents the aggregated results for all days for which data was available (including the standard deviations σ). All parameters have reasonable values. For example, t_{delay}^{cnx} is higher than t_{delay}^{dir}, since an additional time delay due to a connecting flight is more likely to be accepted if the shortest itinerary in the market already represents a connection. The values from Table 4.4 are used as final parameter setting when applying the connection building routine.

4.2.4 Itinerary Market Share Estimation

4.2.4.1 Overview

The direct flights and connecting flights represent the set of itineraries a passenger can choose from. The objective of itinerary market share estimation is to forecast the attraction of each itinerary for a single passenger. The attraction of an itinerary depends on attributes such as convenience of travel, travel time, departure and arrival time, average fare, aircraft type, and airline preferences. This attraction can be interpreted as the market share of the itinerary, thus, if multiplying this value with the market size estimated in Sect. 4.2.2, the total passenger demand for each itincrary can be calculated.

Although there are a number of publications on forecasting techniques and studies of passenger demand forecasting (see Coldren and Koppelman (2005) for an overview), few published models are available which are able to forecast itinerary market shares. The most common model used for this type of estimation is the multinomial logit (MNL) model and its variants; some examples are published in detail including its variables (see for example Coldren and Koppelman (2005) and Bauer (2004)). Because for this study the variables required and used in the published models are not available, the models cannot be used. Instead, a new model has to be developed and calibrated using the data available for this study.

One reason why MNL models are commonly used in forecasting is their well-defined structure leading to easy calibration and fast computation times. However, it remains unclear whether the basic structure of MNL models (like logistic function, linear-in-parameter utility) limits the forecasting accuracy in comparison to other structures or less structured models. Thus, in addition to a traditional MNL model, a custom model for the estimation of itinerary shares (EIS) is developed and calibrated. As the same input data is used for both models, they can be compared well and evaluated according to their forecasting quality. The model with the highest prediction quality is then used as an itinerary market share estimation model in the overall schedule evaluation process.

The next section describes the basic setup including the variables and approach used for calibration and evaluation of both models. Sections 4.2.4.3 and 4.2.4.4 present the MNL model and the EIS model. These sections also give details on the calibration and evaluation of the models. In Sect. 4.2.4.5 the forecasting quality of both models is compared.

4.2.4.2   Setup

The impact of attributes of itineraries on their attraction can be modeled either separately for each city pair, or aggregated for all city pairs. If modeled separately for each city pair, model parameters are different between markets, whereas the aggregated calibration – as conducted here – results in method parameters applicable to all city pairs. Thus, the model can be used for estimation in new markets which is important for flight schedule construction. In addition, this study does not use different passenger segments or time periods resulting in group-specific or time-specific coefficients. Instead, each model is calibrated using all available data.

The total demand in one market is calculated as the sum of all bookings over all itineraries in the market. By dividing the number of passengers on one itinerary by the total number of passengers in the market, the market share of this itinerary can be calculated. By using the market share as the dependent variable, an aggregate

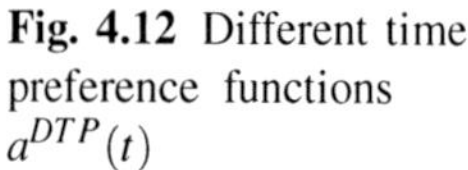

Fig. 4.12 Different time preference functions $a^{DTP}(t)$

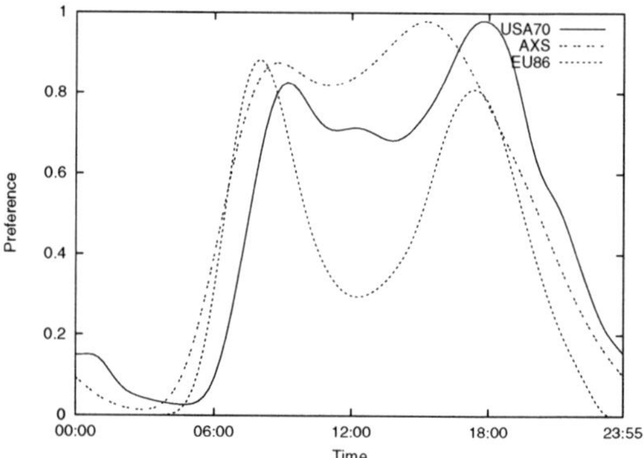

forecasting model for all city pairs can be built and the effects of different market sizes are eliminated.

Data. In this study, the MIDT data described in Sect. 4.1.3 was used. This data measures the realized passenger demand for each itinerary in a market (thus, the market share can be calculated) and provides the corresponding attributes of the itineraries. A market consists of all itineraries available on one day between a pair of airports. Only markets with at least two itineraries are considered in the study (if only one itinerary exists no estimation of the market share is necessary). The resulting data set contains 2,978 different city pairs with a total of 961,430 itineraries.

In principle, the number of attributes of an itinerary can be large depending on the level of detail. Table 4.5 lists the attributes (independent variables) that are used for this study to describe relevant properties of itineraries. It also presents a short description of each variable, its range, and if necessary, the functional form as used in the different models. The different variables are modeled in such a way that the impact of the variable on the attraction of an itinerary increases with higher values.

The variable $a^{DTP}(t)$ requires further explanation. Passengers usually have preferences for specific departure times, thus, time preferences do not stay constant during the day. For example, standard business travelers are likely to prefer departure times in the morning and in the afternoon/evening. $a^{DTP}(t)$ describes how the preference for a specific departure time changes throughout a day. In this study, three different $a^{DTP}(t)$ functions are considered, that are plotted in Fig. 4.12.

- USA70: This function is derived from a survey of domestic airline traffic conducted in 1969 by the US Department of Transportation (O'Connor, 1982).
- AXS: This function is used in a software used by an airline for schedule evaluation.
- EU86: This function is derived from a study in 1986 on passenger volumes on short-haul routes in Europe published by Biermann (1986).

Calibration and Evaluation. The goal of calibration is to adjust the parameters of each forecasting model so that the model reproduces the calibration data in the

Table 4.5 Description of explanatory (independent) variables representing relevant properties of itineraries

Variable	Values	Functional form	Description
Travel time ratio	[0,1]	$a_i^{TTR} = \max\left(2 - \frac{time_i}{time_{sh}}, 0\right)$	Ratio between total travel time $time_i$ of itinerary i and travel time $time_{sh}$ of shortest itinerary sh in the market.
Itinerary type	{0,1}	$a_i^{TYP} = \begin{cases} 1 & \text{if } i \text{ is direct flight} \\ 0 & \text{if } i \text{ is connection} \end{cases}$	Discrete value indicating direct flight or connection.
Shortest itinerary type	{0,1}	$a_i^{STY} = \begin{cases} 1 & \text{if } sh \text{ is connection} \\ 0 & \text{if } sh \text{ is direct flight} \end{cases}$	Discrete value indicating if shortest itinerary sh in the market is direct flight or connection.
Departure time preference	[0,1]	$a^{DTP}(dep_i)$	Indicates the attraction of the departure time dep_i of itinerary i for a potential passenger (see Fig. 4.12).
Airline quality/ preference	[0,1]	a_i^{QUA}	Describes the quality of the airline operating itinerary i as published by Skytrax (2006).
Airline presence	[0,1]	a_i^{PRS}	Indicates the total market share of the airline operating itinerary i in the market.
Closeness (closest itinerary)	[0,144]	$a^{CLO} = 144 - |dep_i - dep_{cl}|$	Time difference between departure time dep_i of itinerary i and departure time dep_{cl} of the closest (with respect to time) itinerary in the market. Time is measured in 5-minute-intervals (maximal time difference is 144 (12 hours)).
Travel time ratio (closest itinerary)	[0,2]	$a_i^{TRC} = 2 - \frac{time_i}{time_{cl}}$	Ratio between total travel time $time_i$ of itinerary i in comparison to travel time $time_{cl}$ of the closest itinerary in the market.

best way. In this study, the process of calibration and evaluation is the same for both models. Out of the total number of observations, a set of randomly chosen itineraries serves either as a calibration data set (CS) or as a validation data set (VS). The CS is used to calibrate each model. Then, the calibrated model is evaluated by measuring the forecasting quality using the data of the VS. The forecasting quality of each model is evaluated using the mean squared error (MSE)

$$MSE = \frac{\sum_k (p_k - t_k)^2}{|K|}, \tag{4.8}$$

where $|K|$ is the number of elements in the total set K of itineraries, p_k is the market share predicted for itinerary $k \in K$, and t_k is the observed market share.

4.2.4.3   Multinomial Logit Model

In this section, a multinomial logit (MNL) model for itinerary market share forecasting is formulated and tested. Multinomial logit models are commonly used methods for *Discrete Choice Problems* in which a person has to choose one alternative from a given set of alternatives. Although MNL models are common in market research and airline planning, only a few publications of MNL models for itinerary market share estimation are available (Coldren & Koppelman, 2005). For examples see Ashford and Benchemam (1987), Alamdari and Black (1992), Coldren et al. (2003), and Hsu and Wen (2003).

Formulation. In the following, a MNL model for the itinerary market share estimation problem of this study is presented. See Train (2003), Ben-Akiva and Lerman (1985), or Kanafani (1983) for more details.

In general, it is assumed that each passenger acts rationally and wants to maximize his utility (Ben-Akiva & Bierlaire, 1999). The utility or value V_k of a given itinerary $k \in K$ for a passenger $n \in N$ depends on the attributes $a \in A$ of the itinerary. In MNL models, V_k is a linear combination of the attribute values $\mathbf{X_k} = (x_{k1}, x_{k2}, \dots, x_{ka})$ and method parameters $\beta = (\beta_1, \beta_2, \dots, \beta_a)$:

$$V_k = \beta^T \mathbf{X_k} \tag{4.9}$$
$$= \beta_1 x_{k1} + \beta_2 x_{k2} + \beta_3 x_{k3} \dots \beta_a x_{ka}. \tag{4.10}$$

In this formulation, the value of one itinerary depends only on the characteristics of this alternative. Attributes of individual passengers are not included in the model and all passengers are grouped to one single entity with the same value perception. Furthermore, one vector β is assumed for the total model without segmenting the total number of observations to build individual parameters.

The probability p_k (attraction) of an itinerary k to be chosen by one passenger $n \in N$ is defined by:

$$p_k = \frac{e^{V_k}}{\sum\limits_{k} e^{V_k}}. \tag{4.11}$$

During calibration, the objective is to determine the vector $\hat{\beta}$ that maximizes the likelihood of the observation. y_{nk} is defined as:

$$y_{nk} = \begin{cases} 1 & \text{if individual } n \text{ chooses alternative } k \\ 0 & \text{otherwise.} \end{cases} \tag{4.12}$$

Then, the probability of the choice of individual n is given by

$$\prod\limits_{k} (p_k)^{y_{nk}}. \tag{4.13}$$

Because $y_{nk} = 0$ for all non-chosen alternatives, this term is simply the probability p_k of the chosen alternative k.

Since the individual choices are independent, the probability of the correct prediction of all N individual choices is given by the likelihood function:

$$L(\beta) = \prod_{n=1}^{N} \prod_{k} (p_k)^{y_{nk}}. \tag{4.14}$$

The relative attraction R_k of an itinerary k is calculated using the number of passengers D_k that choose alternative k and D as the total number of passengers:

$$R_k = \frac{D_k}{D} \tag{4.15}$$

with

$$D_k = \sum_{n=1}^{N} y_{nk} \tag{4.16}$$

and

$$D = \sum_{k} D_k. \tag{4.17}$$

Because the choices of different individuals are assumed to be independent and identically distributed (Bernoulli trials), the joint probability is given by the multinomial distribution. Therefore, the likelihood function can be calculated as:

$$L(\beta) = \frac{D!}{D_1! D_2! \dots D_k!} \prod_{k} (p_k)^{D_k}. \tag{4.18}$$

This likelihood function applies to a given set of competing itineraries. When considering different city pairs and days in model calibration, itineraries have to be separated into individual groups with competition within but not between each other. By defining a market as the combination of a city pair and a day, M as a set of markets, $m \in M$ as one market, and K_m as the set of itineraries competing in one market m, the likelihood function is given as:

$$L(\beta) = \prod_{m=1}^{M} \frac{D_m!}{\prod_{k \in K_m} D_k!} \prod_{k \in K_m} (p_k)^{D_k}. \tag{4.19}$$

The maximum likelihood estimator (MLE) is the value $\hat{\beta}$ that maximizes this function. Taking the logarithms simplifies maximization resulting in the log-likelihood function:

$$LL(\beta) = \sum_{m=1}^{M} \left\{ \ln D_m! - \sum_{k \in K_m} \ln D_k! + \sum_{k \in K_m} D_k \ln p_k \right\}. \tag{4.20}$$

$LL(\beta)$ is globally concave with respect to β simplifying the estimation of $\hat{\beta}$ (Mc-Fadden, 1974).

To test the overall model (structure), the log-likelihood ratio index ρ^2 is computed as:

$$\rho^2 = 1 - \frac{LL(\hat{\beta})}{LL(0)} \tag{4.21}$$

with $LL(\hat{\beta})$ as the value of the log-likelihood function for the estimated value $\hat{\beta}$ and $LL(0)$ as the log-likelihood value for $\beta = 0$. Because additional independent variables never reduce ρ^2, the log-likelihood ratio is corrected by the number of variables A:

$$\bar{\rho}^2 = 1 - \frac{LL(\hat{\beta}) - A}{LL(0)}. \tag{4.22}$$

Although having the same objective as the coefficient of determination R^2 in classical linear regression, the interpretation of $\bar{\rho}^2$ is not exactly the same. Values of $\bar{\rho}^2$ between 0.2 and 0.4 are assumed to indicate an acceptable model fit (Urban, 1993).

To determine the significance of individual variables, standard t-tests can be used (Train, 2003). For each β_a the null hypothesis H_0 is tested against the alternative hypothesis H_α:

$$H_0 : \beta_a = 0 \tag{4.23}$$

$$H_\alpha : \beta_a \neq 0. \tag{4.24}$$

Calibration and Validation. In model calibration, the objective is to find $\hat{\beta}$ which is conducted by Maximum-Likelihood-Estimation. All variables described in Table 4.5 were included, however, there is a choice between the three different time preference functions. In a set of experiments, the impact of the different time preferences is analyzed. Using each time preference function separately, an MNL model is calibrated and validated in 10 experiments, each with a different set of 40,000 randomly chosen itineraries as CS and VS. Fig. 4.13 shows the averages of the LL-ratio and MSE (of the VS) for each time preference function. The consideration of the

Fig. 4.13 Results of time preference function selection for the MNL model

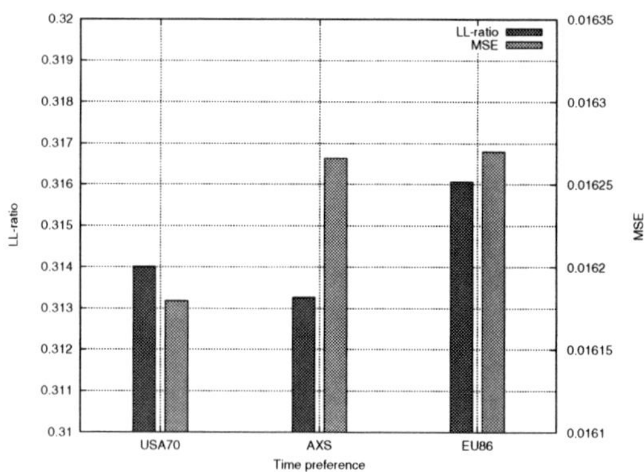

different time preference functions yields similar results, and the LL-ratios between 0.313 and 0.317 indicate an acceptable model fit. Because USA70 has the lowest MSE, this time preference function should be used within this MNL model for itinerary share estimation.

4.2.4.4 Custom Model

In this section, a custom model (denoted as EIS model) for the estimation of itinerary market shares is presented. Analogous to the MNL model, the same data is used to calibrate the EIS model and to compare its performance to the MNL model.

Model Description. The attraction $A_k(t)$ of an itinerary $k \in K$ at time $t \in [t_{min}, t_{max}]$ is estimated based on its attributes $\mathbf{X_k(t)} = (x_{k1}, x_{k2} \ldots, x_{ka-1}, x_{ka}(t))$ and the parameters $\beta = (\beta_1, \beta_2 \ldots, \beta_a)$. The attribute $x_{ka}(t)$ depends on the time-dependent departure time preference $a^{DTP}(t)$. Appropriate values for the parameters β are determined in the calibration phase.

The attraction of an itinerary k is calculated by summing up the weighted attributes:

$$A_k(t) = \beta^T \mathbf{X_k(t)} \tag{4.25}$$

$$= \beta_1 x_{k1} + \beta_2 x_{k2} + \beta_3 x_{k3} \ldots \beta_{a-1} x_{ka-1} + \beta_a x_{ka}(t). \tag{4.26}$$

The departure time preference $a^{DTP}(t)$ of an itinerary k is weighted by the difference between the departure time dep_k of itinerary k and the preferred departure time t of a passenger. With increasing difference $|dep_k - t|$ between departure time dep_k and preferred departure time t of a passenger, $x_k(t)$ follows a Gaussian function (Douglas & Miller, 1974). The attraction of itinerary k is maximal for all passengers who want to fly at time $t = dep_k$. It is calculated as

$$x_k(t) = a^{DTP}(t) \exp(-(t - dep_k)^2 / \lambda), \tag{4.27}$$

where the different possibilities for $a^{DTP}(t)$ are shown in Fig. 4.12 and λ is a free parameter that is adjusted during model calibration. The assumed distribution of the preference might be different for each passenger (segment). Dobson and Lederer (1993) show that the preference function depends on the purpose of travel. For example, business travellers need to arrive at the destination on time and, thus, would accept earlier flights compared to later flights. If additional information on this behavior was available, alternative preference functions could be tested and implemented.

$S_k(t)$ is defined as the absolute, normalized attraction ($S_k(t) \in [0, 1]$) of itinerary k for any time $t \in \{t_{min}, t_{max}\}$. It is calculated as

$$S_k(t) = \frac{A_k(t)}{\max_{i \in K_m} \max_{t \in \{t_{min}, t_{max}\}} (A_i(t))}, \tag{4.28}$$

where K_m is the set of itineraries in the market m that itinerary k belongs to ($k \in K_m$). Therefore, the denominator finds the maximal attraction over the time horizon of all itineraries that belong to the same market K_m as itinerary k. $S_k(t)$ represents the attraction of itinerary k independent of competing itineraries.

If there is more than one itinerary k in one market K_m, a potential passenger can choose between the competing itineraries. The relative attraction $R_k(t)$ of an itinerary k is given as

$$R_k(t) = \frac{S_k(t)}{\max\left(\sum_{i \in K_m} S_i(t), 1\right)}.$$ (4.29)

It is assumed that time is discretized and $t \in \{t_{min}, t_{max}\}$.

Calibration and Validation. The goal of calibration is to find the parameters $(\beta_1, \beta_2 \ldots, \beta_a)$ and λ that yield the highest forecasting quality (minimum MSE). For this purpose a simple hill climbing metaheuristic is used. This procedure is equivalent to algorithm 3 on page 84. A solution is encoded as a vector of parameters (consisting of $(\beta_1, \beta_2 \ldots, \beta_a)$ and λ). The quality of a solution is the MSE of the calibration set. Thus, a higher MSE implicates a lower solution quality. The initial solution is created randomly. A local search operator modifies a randomly chosen parameter β towards β^* by adding a random value selected according to a normal distribution $N(0, 0.25)$ with mean $\mu = 0$ and variance $\sigma^2 = 0.25$.

All variables from Table 4.5 are used for the EIS except a^{CLO} and a^{TRC}. These variables are excluded because their effects are implicitly included in the model structure in which the attraction is calculated for each point in time taking into account the competing itineraries' attractions.

Equation (4.25) calculates the attraction of an itinerary as the sum of its weighted attributes. Two other, alternative, model formulations were tested. In the first alternative model formulation (*MultAll*), addition was replaced by multiplication. In the second alternative model formulation (*MultTime*), the attraction of an itinerary k is calculated as

$$A_k(t) = (\beta_1 x_{k1} + \ldots + \beta_{a-1} x_{ka-1}) \beta_a x_{ka}.$$ (4.30)

Fig. 4.14 shows the resulting MSE over different time preference functions for the three different model formulations. Ten experiments were performed with a different set of 40,000 randomly chosen itineraries as CS and VS for each alternative. The results indicate that the additive model (denoted as *Add*) described in (4.25) yields the best results as for all three time preference functions the MSE is lower than for the alternative models. The MSE is lowest when using EU86 as the time function.

4.2.4.5   Evaluation

Because both models examined in this study have the same objective and require the same input data, a comparison between them is straightforward. In the previous sections, the best setting for each model was determined. This section compares the resulting models in an identical experimental setup. In this setup, the CS contains 50,000 randomly chosen itineraries. For validation, ten independent runs with

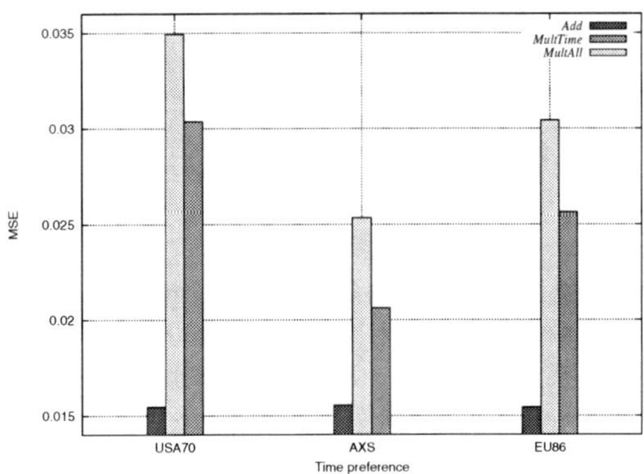

Fig. 4.14 Results of time preference function selection for different EIS models

50,000 randomly chosen itineraries were performed. Each experiment is repeated ten times with randomly chosen data sets.

Table 4.6 presents the parameter estimates and MSE for the MNL and EIS model.

Table 4.6 Comparison of the results of both itinerary market share models

Variable	MNL	EIS
a^{TTR}	0.337	0.392
a^{TYP}	1.996	1.550
a^{STY}	-0.009	1.218
a^{DTP}	0.125	0.827
a^{QUA}	0.009	0.062
a^{PRS}	0.720	0.502
a^{TRC}	0.055	
a^{CLO}	0.003	
λ		5.937
$\bar{\rho}^2$	0.333	
MSE	0.0165	0.0156

The results of the MNL model show a high log-likelihood ratio index $\bar{\rho}^2$, indicating a valid model structure and good fit. All variables are significant on the 99.9%-level in most experiments. In one experiment, a^{QUA} is significant on the 90%-level, a^{STY} is significant on the 75%-level in three experiments, and two other experiments resulted in a^{STY} becoming insignificant.

t-tests on the parameters of the EIS model show significance on the 99.%-level for all variables and experiments, except a^{QUA} which is significant on the 95%-level for one experiment.

Because of the different model specification, a direct comparison of the parameter values between the MNL and EIS model is meaningless. However, when looking at each model separately, the estimates of the different variables can be interpreted with respect to their impact on an itinerary's attraction or market share. In particular, the estimates for the travel time ratio (a^{TTR}) indicate a positive impact of a shorter travel time on the attraction of an itinerary. A lower travel time (representing an increase of the variable) results in an increase of the attraction of an itinerary. The same effect can be observed for the itinerary type (a^{TYP}). Passengers prefer direct flights and avoid connection flights due to the increased travel time, the inconvenience of switching planes, or higher probability of delays and lost baggage. This has also been observed by Coldren et al. (2003). This effect is especially strong because the used data sets contain only short-haul routes (itineraries between Germany and European countries). When applying the model to long-haul routes, the advantage of direct flights is expected to be lower due to the reduced perceived disadvantage of connection flights in comparison to direct flights. A high positive impact on attraction can also be observed for an airline's presence in a market (a^{PRS}). This reflects the strong position of national air carriers on routes to or from their home countries. Usually, a carrier with a high presence in a market can offer more flights and get more acknowledgment from potential passengers. This was also observed by Teodorovic and Krcmar-Nozic (1989). Furthermore, the results for a^{DTP} indicate that passengers prefer departure times following the time preference function assumed. The impact of an airline's quality is low (a^{QUA}). This can be explained by the low differences of service qualities between airlines offering air service in Europe. In addition, flights are short and quality is only an important factor for long-haul flights. Finally, all variables representing the competition in a market (a^{STY}, a^{TRC}, and a^{CLO}) have a low impact on the attraction of an itinerary.

Comparing the MSE of the two different models indicates that the EIS model outperforms the MNL. To confirm these observations, an unpaired t-test on the previous results is conducted. The null hypothesis H_0 is that the observed differences in the forecasting quality (EIS outperforms MNL) are random. H_α says that the differences are a result of the model specification. The critical t-value for $p = 0.999$ is 3.6105. The t-value for a comparison between the results of the MNL and EIS model is 18.9338, thus, H_0 can be rejected on the 99.9%-level.

4.2.4.6 Final Model

Although the MNL is the standard model for market share estimation used by airlines, it is outperformed by the EIS model in this study. Therefore, this model (with EU86 as time preference function) is used as the market share estimation step within the overall schedule evaluation process. Applying the complete data set to this model leads to the following final set of parameters used in the EIS for schedule evaluation:

$$a^{TTR} = 0.287,$$
$$a^{TYP} = 1.167,$$
$$a^{STY} = 0.937,$$
$$a^{DTP} = 0.767,$$
$$a^{QUA} = 0.049,$$
$$a^{PRS} = 0.359,$$
$$\lambda = 5.763.$$

4.2.5 *Passenger Allocation*

Given the market sizes and the relative share (attraction) of each itinerary in the market, calculating the absolute passenger demand for each itinerary is straightforward. This demand competes for the limited capacity of the aircraft of the single flights the itinerary are constructed of. Because each flight might be a leg of a connecting itinerary, this competition takes place between itineraries of different markets. The objective of the passenger allocation step is to satisfy the passenger demand by providing an assignment of this demand to the itineraries (and its flights, respectively) without violating capacity constraints (Mathaisel, 1997). This task is commonly referred to as *spill & recapture*.

Since this problem is of minor scientific interest, except for the model of Mathaisel (1997) no applicable publication on this topic could be found. The model of Mathaisel (1997) iteratively assigns fractions of the total demand to the available flights of the demanded itineraries, imitating the booking behavior of passengers. Each iteration can be interpreted as one day on which bookings take place. As an aircraft is filled with passengers, the probability to get a place in an aircraft decreases or the fare increases, reducing the attraction of the corresponding itinerary. Thus, after each simulated day of booking the attraction of each itinerary has to be recalculated under consideration of available capacities, considerably increasing the total computation time of this approach.[9] Therefore, although the model of Mathaisel (1997) better reflects the real passenger allocation process, in this study another procedure is used. In general, it consists of three steps. In a first step, the total demand is assigned to the itineraries regardless of their (limited) capacities. Then, if the assigned passenger number exceeds the capacity of any flight, this exceeding demand is spilled. In a final step, the spilled passenger demand is assigned to other (less desirable) itineraries in the market with free capacities left.

In the following, the procedure of passenger allocation including spill and recapture is presented in detail.

1. The absolute passenger demand D_k^{iti} for each itinerary k is calculated as:

$$D_k^{iti} = s_m \cdot R_k \tag{4.31}$$

   with s_m as the market size of market m, R_k as the relative attraction (see Eq. 4.29 on page 92) of itinerary $k \in K_m$.

2. The passenger demand D_f^{flight} for flight $f \in F$ is calculated by summing up the passenger demand D_k^{iti} of all itineraries of all markets that use this flight (either as direct flight or as leg of a connection):

$$D_f^{flight} = \sum_{\forall\, k \text{ using } f} D_k^{iti} \tag{4.32}$$

3. For each flight f, the rate of the exceeding demand e_f is calculated as:

---

[9] In Fig. 4.15 computation times for the calculation of attractions (depending on the number of flights or itineraries, resp.) are presented.

$$e_f = D_f^{flight} / c_f \tag{4.33}$$

with c_f as the capacity of the fleet type of f. c_f could be multiplied by the observed average seat load factor. Because of the effect of day-to-day statistical variations in total demand, the average load factor never reaches 100% (Mathaisel, 1997).

4. The rate of exceeding demand on the itinerary level e_k depends on the highest rate of its flights:

$$e_k = max(e_f) \ \forall \ f \text{ used by } k \tag{4.34}$$

5. Based on the given demand, the number of allocated passengers P_k^{iti} to the itinerary and P_f^{flight} to the flight can be calculated as:

$$P_k^{iti} = D_k^{iti} / e_k \ \forall \ k \text{ with } e_k > 1 \tag{4.35}$$

$$P_f^{flight} = \sum_{\forall \ k \text{ using } f} P_k^{iti} \tag{4.36}$$

6. Passenger demand not allocated to itineraries and flights is spilled and given on an itinerary level (sP_k^{iti}) and market level (sP_m^{market}):

$$sP_k^{iti} = D_k^{iti} - P_k^{iti} \tag{4.37}$$

$$sP_m^{market} = \sum_{k \in K_m} sP_k^{iti} \tag{4.38}$$

7. The free capacity fc^{flight} of flight f and fc^{iti} of itinerary k after passenger allocation is:

$$fc_f^{flight} = c_f - sP_f^{flight} \tag{4.39}$$

$$fc_k^{iti} = min\left(fc_f^{flight}\right) \ \forall \ f \text{ used by } k \tag{4.40}$$

8. Any itinerary with free capacities left might recapture spilled demand. The set of itineraries with free capacities is denoted as $J \subseteq K$:

$$J = \{k | k \in K, fc_k^{iti} > 0\}. \tag{4.41}$$

9. Relative attractions R^* of all itineraries $j \in J$ are recalculated and normalized:

$$R_j^* = \frac{R_j}{max\left(\sum_j R_j, 1\right)}. \tag{4.42}$$

10. Finally, the passengers' recapture is once calculated following steps 1-6 and using J as set of itineraries, R^* as the attraction of all $j \in J$, sP_m^{market} as the remaining market size of market m, and fc_f^{flight} as remaining capacities of the flights $f \in F$.

4.2.6 Profit Estimation

Given the number of passengers on the itineraries, the calculation of the overall profit of a flight schedule F is straightforward. Let c_f^{block} denote the block hour costs of the aircraft type (see Sect. 4.1.3) assigned to flight f, t_f^{block} the block time of flight f, y_m the passenger yield in market m, and P_k the number of passengers on itinerary $k \in K$. The operating profit π_F of flight schedule F is calculated as:

$$\pi_F = \sum_{m} \sum_{k \in K_m} P_k \cdot y_m - \sum_{f \in F} c_f^{block} \cdot t_f^{block} \qquad (4.43)$$

To calculate the profit π_f of a single flight f, the passenger yield of itinerary connections has to be distributed to the connecting flights. The portion of the profit assigned to a single flight is proportional to its share of total block time t_k^{block} of the connection:

$$\pi_f = \sum_{\forall \, k \text{ using } f} P_k \cdot y_m \cdot \frac{t_f^{block}}{t_k^{block}} - c_f^{block} \cdot t_f^{block} \qquad (4.44)$$

4.2.7 Summary

One important piece of information when creating and optimizing an airline schedule is its quality. The integrated scheduling approaches presented in this study represent iterative search procedures in which one or multiple schedule evaluations have to be conducted in each iteration. Airlines have access to sophisticated evaluation tools that are able to predict passenger flows and a revenue and cost assignment on a detailed level based on many different variables and factors. However, these tools and the required data usually represent classified information not accessible for this study. In addition, because of the high level of detail, tools used by airlines usually have high computation times lasting from minutes to hours. Therefore, the development of a schedule evaluation procedure is necessary that can be applied in the integrated models presented in this study.

Each schedule is evaluated according to its operating profit following a five-step process:

1. *Market Size Estimation*: The objective is to estimate the total number of air passenger demand between any two airports. In this study, a gravity model to accomplish this task was developed and calibrated using the ordinary-least-squares method with the given data. This model uses geo-economic variables as input to estimate passenger volumes, thus, it can be applied to markets in which flight services have not existed yet.
2. *Itinerary Construction*: Every flight represents a direct travel itinerary between the corresponding airports and might be a leg of a connecting itinerary. The objective of this step is to construct feasible and reasonable connecting itineraries. A set of rules including parameters was defined. The parameters were calibrated with the given data using a simple hill-climbing metaheuristic.

3. *Itinerary Market Share Estimation*: In this step, the individual attraction and, thus, market size of each itinerary is estimated. MNL models are an established technique for this purpose, however, a custom model developed in this study produced better estimates (a comparison using identical experimental setups was conducted). The parameters of the custom model were calibrated using a simple hill-climbing algorithm.

4. *Passenger Allocation*: Given the total demand for travel itineraries, the passengers are allocated to the flights under consideration of capacity constraints of the aircraft. In this study, first passengers are allocated regardless of capacities. Then, the number of spilled passengers is calculated. In a final step, the spilled passengers are recaptured by flights with capacities left.

5. *Profit Estimation*: In this step, the profit is estimated using the passenger numbers, the yield for each market, and operating costs.

The objective of the market size estimation is the prediction of the number of passengers that want to travel with an airline between any two cities. Because this number is independent of a current schedule, this step is only necessary once before the construction and optimization of an airline schedule is conducted. The other evaluation steps depend on the schedule under investigation. The number of travel alternatives then determine the complexity of the schedule evaluation (Belobaba, 1987). As an example for the increasing complexity, Fig. 4.15 presents the number of itineraries constructed depending on the number of flights in the schedule. In addition, the corresponding computation times for the itinerary construction

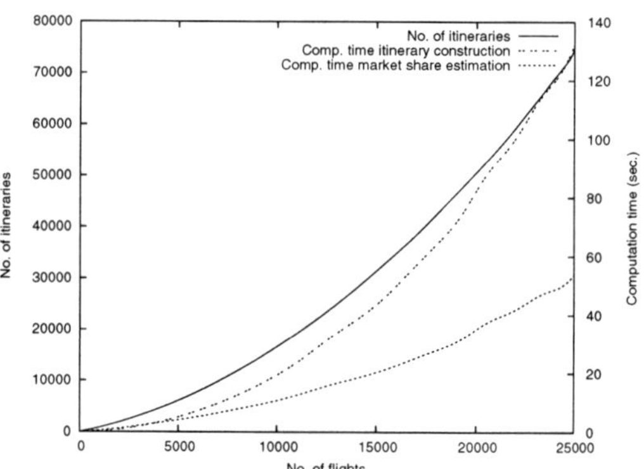

Fig. 4.15 Schedule evaluation computation times

and market share estimation are plotted.[10] An exponential growth of the number of itineraries is clearly visible. In addition, since the effort to calculate market shares directly depends on the number of itineraries, the required computation time also

---

[10] The flights are chosen randomly; for each schedule size, ten evaluation runs are conducted (with the completely calibrated evaluation procedure) on a workstation with an Intel® Pentium® 2.80 GHz processor and 1.00 GB RAM. The results are averaged over ten experiments. The number of routes (see page 79) for the calibrated model and the given data is 4,452,584 for 320 airports.

grows exponentially. Presenting the computation times also gives an impression of the total effort to construct and optimize airline schedules, since the schedule evaluation represents the computationally intensive part of both solution approaches presented in this study.[11]

The presented schedule evaluation procedure had to be developed due to the lack of sufficient information of established evaluation techniques (or rather the availability of commercial applications) and the unavailability of data that could be used directly. Because each single step presented in the previous section relies on the information available for this study, there is room for further enhancements. In general, if more historical data was available for different market segments, on a higher level of detail, or for more time periods, the parameters of each step could be calibrated more specifically. For example, multiple gravity models for different markets could be developed, replacing the aggregate model used in this study. Parameters for the itinerary construction step could be defined specifically for each market. Additional data could also help in constructing different model specifications for the market size or itinerary market share estimation step with possibly better forecasts.

4.3 Sequential Approach

4.3.1 Overview

In this section, a sequential and iterative approach for the development of a complete airline schedule is presented. It includes the flight schedule generation and aircraft scheduling problems from Fig. 2.2. The objective is to develop an airline schedule that is feasible and has a high operating profit. The approach represents an iterative procedure in which in each iteration all subproblems are solved in a sequence until the operating profit does not improve in further iterations, the then current schedule is expected to have maximum profit. An overview of the sequential approach is presented in Fig. 4.16.

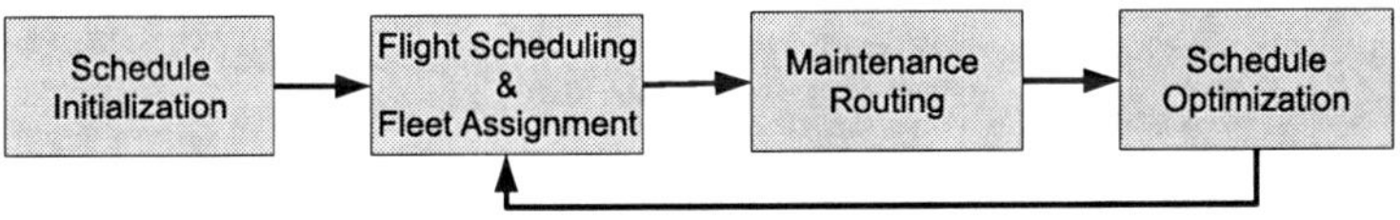

Fig. 4.16 Sequential planning approach

The solution steps *Fleet Assignment, Flight Scheduling,* and *Maintenance Routing* are solved using models from literature, the last step *Schedule Optimization* consists of three algorithms developed in this study. Each step has individual restrictions on its input data and might be unable to create feasible solutions for the

---

[11] Thus, when presenting results of experiments of the airline scheduling approaches, only the number of schedule evaluations necessary is presented instead of always presenting computation times.

given input. To assist each step in finding a solution and to model the interdependencies between the steps, additional *supportive functions* are necessary and applied between the individual steps.

In the following section, the solution steps are presented in more detail including some modifications necessary for their application in the overall scheduling approach. Then, in Sect. 4.3.3 these solution steps are put together including the supportive functions yielding the entire integrated approach for airline scheduling. As illustrated in Fig. 4.16, it is assumed that an initial schedule is given. Because the initialization of a schedule as conducted in this approach builds on elements of the solution steps and supportive functions, this process is described after the presentation of the sequential optimization approach in Sect. 4.3.3.2. Sect. 4.3.4 presents the application of the integrated approach to different planning scenarios. For this purpose, some parameters have to be chosen. In Sect. 4.3.4.1 the calibration of the parameters is described. Then, Sect. 4.3.4.2 presents an analysis of the search process and the resulting solutions of the complete procedure. Finally, a summary is given in Sect. 4.3.5.

4.3.2 Solution Steps

In this section, the main steps shown in Fig. 4.16 are presented. For the fleet assignment and flight scheduling problem, a model of Rexing et al. (2000) combining these two problems is used. The maintenance routing problem is solved following the model by Gopalan and Talluri (1998a). Both models are presented in the following, each followed by a description of some modifications and enhancements to improve their application in the integrated airline scheduling process. Finally, the schedule optimization step consisting of three different algorithms is presented.

4.3.2.1 Fleet Assignment and Flight Scheduling

Model. The fleet assignment and flight scheduling is conducted following the model by Rexing et al. (2000). This model is based on the approach by Hane et al. (1995) and integrates the daily fleet assignment with the flight scheduling into one single model. Given a flight schedule, a time interval around the actual departure time of each flight is specified. The model is then allowed to change the departure time in the interval and assigns a fleet type. Within each time interval, demand is assumed to be constant and independent of the actual departure time and fleet assigned. If the time window size is small (for example 30 minutes), changes in the departure time will be small, limiting demand variations. The objective is to minimize the direct operating costs and opportunity costs. Opportunity costs exist if potential passengers must be left behind because the demand for a flight is higher than the assigned fleet's capacity. The general constraints of the fleet assignment problem presented on page 18 are included in this model.

The fleet assignment problem is represented as a network flow problem consisting of nodes, ground arcs and flight arcs. Each fleet type is represented by a separate network containing all airports in consideration. The nodes represent departures or arrivals at an airport at a specific time. The nodes at each airport are ordered by their time, thus every node has a direct predecessor and a direct successor. A time line is constructed by connecting these nodes by ground arcs, representing an aircraft resting on the ground between its arrival and subsequent departure. Flights are represented as flight arcs connecting the corresponding nodes. The arrival node of each flight arc is placed at the ready time (block time + turn time) of the flight. To guarantee flow balance in the daily fleet assignment, ground arcs connect the last and the first node at every airport. The number of aircraft in use can be calculated by summing up all flows on all (flight and ground) arcs at one point in time (*count time*). Fig. 4.17 illustrates one example of a flight network with two airports.

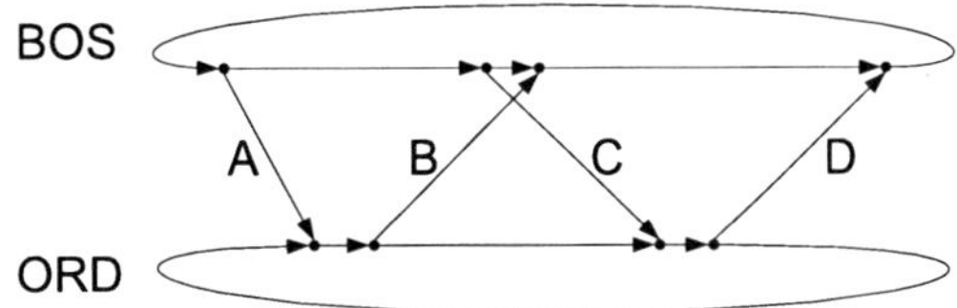

Fig. 4.17 Two-airport flight network (Source: Rexing et al. (2000))

By introducing time windows for each flight, the flight scheduling problem can be included into this representation of the problem. A time window specifies how much the original departure time is allowed to vary. The additional flexibility in choosing departure times is expected to result in a more efficient (less expensive) schedule. For example, two flights are assigned to two different aircraft because the ready time of the first flight is later than the departure time of the second flight. If the departure time (and thus arrival/ready time) of the first flight can be shifted forward and the departure time of the second flight can be delayed, it might become possible to assign both flights to the same aircraft. In this model formulation, the time window is split up in intervals (for example five-minute intervals) and copies of the original flight (arc) are placed at every interval, representing alternative departure times (see Fig. 4.18). The model is then allowed to choose only one copy of each set, fixing

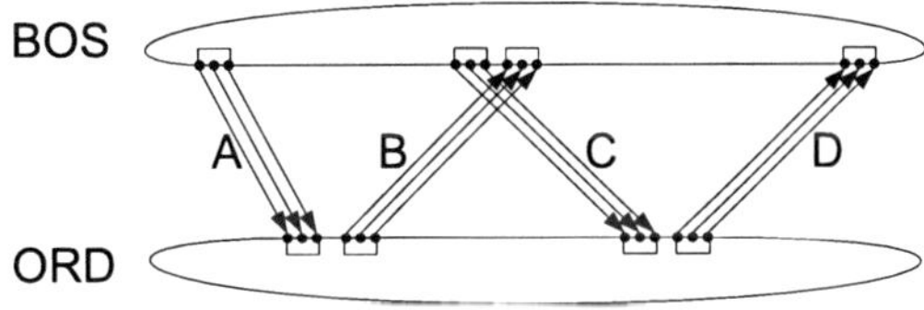

Fig. 4.18 Two-airport flight network with time windows (Source: Rexing et al. (2000))

the final departure time for each flight. Introducing time windows is quite simple, however, the resulting model becomes computationally expensive. One parameter to control the difficulty of the problem is the number of copies per flight and the time window size. Because it is assumed that there is no variation of the demand within each time window, the size of each window should be kept small to comply with this assumption, however, limiting the degree of freedom for flight scheduling in this combined model.

After constructing the flight networks for each fleet, the fleet assignment problem can be solved using the following formulation.

Minimize:

$$\sum_{k \in K} \sum_{i \in F} \sum_{n \in N_{ik}} c_{ik} x_{nik} \tag{4.45}$$

Subject to:

$$\sum_{k \in K} \sum_{n \in N_{ik}} x_{nik} = 1 \quad \forall i \in F \tag{4.46}$$

$$\sum_{k \in K} \sum_{n \in N_{ik}} b1_{lnik} n_{nik} + \sum_{g \in G_k} b2_{lgk} y_{gk} = 0 \quad \forall l \in L_k, \forall k \in K \tag{4.47}$$

$$\sum_{i \in F} \sum_{n \in N_{ik}} d1_{nik} x_{nik} + \sum_{g \in G_k} d2_{gk} y_{gk} \leq S_k \quad \forall k \in K \tag{4.48}$$

$$y_{gk} \geq 0 \quad \forall g \in G_k, \forall k \in K \tag{4.49}$$

$$x_{nik} \in \{0, 1\} \quad \forall i \in F, \forall n \in N_{ik}, \tag{4.50}$$
$$\forall k \in K$$

Parameters:

F = set of flights
K = set of fleets
S_k = number of aircraft of fleet k
G_k = set of ground arcs in fleet k's network
L_k = set of nodes in fleet k's network
N_{ik} = set of arc copies of flight i in fleet k's network
$|N_{ik}|$ = number of arc copies of flight i with fleet type k
c_{ik} = cost to fly flight i with fleet type k

$$b1_{lnik} = \begin{cases} 1 & \text{if copy } n \text{ of flight } i \text{ begins at node } l \text{ in fleet } k\text{'s network} \\ -1 & \text{if copy } n \text{ of flight } i \text{ ends at node } l \text{ in fleet } k\text{'s network} \\ 0 & \text{otherwise} \end{cases}$$

$$b2_{lgk} = \begin{cases} 1 & \text{if ground arc } g \text{ begins at node } l \text{ in fleet } k\text{'s network} \\ -1 & \text{if ground arc } g \text{ ends at node } l \text{ in fleet } k\text{'s network} \\ 0 & \text{otherwise} \end{cases}$$

$$d1_{nik} = \begin{cases} 1 & \text{if copy } n \text{ of flight } i \text{ crosses the count time in fleet } k\text{'s network} \\ 0 & \text{otherwise} \end{cases}$$

$$d2_{gk} = \begin{cases} 1 & \text{if ground arc } g \text{ crosses the count time in fleet } k\text{'s network} \\ 0 & \text{otherwise} \end{cases}$$

Decision Variables:

$$x_{nik} = \begin{cases} 1 & \text{if copy } n \text{ of flight } i \text{ is flown by fleet } k \\ 0 & \text{otherwise} \end{cases}$$

y_{gk} = number of aircraft on ground arc g in fleet k's network

Equation 4.45 is the objective function minimizing the costs of assigning aircraft types to the flight arcs (operating costs and opportunity costs). Constraint 4.46 requires that each flight is covered by exactly one fleet by allowing only one copy of flight arcs to be chosen. Constraint 4.47 ensures the flow balance at each node, constraint 4.48 limits the number of the available aircraft.

Application. The model was implemented in the overall sequential planning approach as presented in the previous section. Network preprocessing steps suggested by Rexing et al. (2000) to prune the problem before constructing and solving the LP matrix were not implemented, because preliminary tests showed that these steps are not necessary for the problem instances examined in this study (Barth, 2005). In addition, the optimization steps of the integrated approach represent the computationally demanding part during application.

The fleet assignment problem cannot be solved if the number of available aircraft is not sufficient to perform all flights given. Thus, it is necessary to remove flights from the schedule. Therefore, an additional attribute *optional* is introduced and assigned to every flight that might be removed from the schedule. If a flight is optional, constraint 4.46 is changed for this flight in order to allow that no copy of flight arcs at all is covered, resulting in the deletion of the flight from the schedule. Because the fleet assignment model tries to minimize the costs, usually an optional flight is removed from the schedule if no other constraint (for example flow balance) is violated. Furthermore, to meet maintenance restrictions sometimes it is necessary to assign certain flights to specific fleet types and to prevent the presented model from changing this assignment. For example, one fleet type is assigned only to flights that do not depart or arrive at a suitable maintenance station for this type. In such cases, the attribute *maintenance* assigned to those flights indicates that copies of the flight arc are allowed with the fleet type assigned in the previous planning cycle resulting in a similar fleet assignment. Both attributes *optional* and *maintenance* are set outside of the presented flight scheduling and fleet assignment model by the supportive functions presented later (see page 113).

Because Rexing et al. (2000) show that narrow departure times for all copies of one flight cause an extreme increase in the problem size, often without the benefit of providing a substantially better solution than broader departure times would, the number of flight arc copies in this approach is limited to five. The time window size in minutes is set by the parameter tw with all copies of flight arcs evenly distributed within this interval. Flight arcs that violate airport operating hours or curfew restrictions are not included. Following the suggestion in the paper written by Rexing et al. (2000), the count time is set to a time which is crossed usually only by wrap around arcs and a few flight arcs. In this approach, 03:00 a.m. is used as count time.

The costs of each flight (arc) consist of operating and opportunity costs. Operating costs include expenses directly related to the flight like costs for fuel, maintenance, landing fees etc. and can be easily obtained. For this application, the block hour costs multiplied by the block times are used (see page 66). Opportunity or spill costs are calculated by multiplying the number of spilled passengers with the fare they would have paid. This fare or yield is assumed to be given (see page 65). The number of spilled passengers is calculated by subtracting the assigned fleet type's capacity from the unconstrained demand. In this application, the unconstrained demand is calculated using the schedule evaluation model presented in Sect. 4.2 without the spill and recapture step. Thus, the number of spilled passengers is the total of passengers demanding the flight as nonstop itinerary and as part of a connecting itinerary subtracting the capacity of the fleet type of the current flight arc. As a consequence, the connectivity of the flight schedule is incorporated in the opportunity costs.

4.3.2.2 Aircraft Maintenance Routing

Model. The objective of this step is to construct feasible routings for individual aircraft based on an airline schedule and a fleet assignment. To be feasible, the resulting schedule must allow each aircraft to undergo required maintenance at the necessary intervals.

In this planning approach, the aircraft maintenance routing problem is solved following the (infinite-horizon) model of Gopalan and Talluri (1998a). In their approach it is assumed that each aircraft has to undergo maintenance at least once every three days, that maintenance is scheduled at night, and that each aircraft has to undergo one balance check (a more complex maintenance check) during the planning cycle. Furthermore, each fleet type has only one rotation, thus, if n aircraft of one type exist, each aircraft comes back to the starting airport of its rotation after n days and undergoes the balance check every n days.

Because a valid fleet assignment is given by the previous step, the aircraft routing problem can be conducted for each fleet type separately. The planning is performed on the basis of each aircraft's routing or lines-of-flying (LOF). An LOF contains all flights of one aircraft during the day, it can be defined by the first departure airport of the day, the last arrival airport, and the assigned fleet type. The last airport might be a maintenance station at which the aircraft could be checked before starting the LOF of the next day. LOFs are constructed by combining the single flights with simple rules like last-in-first-out (LIFO) or first-in-first-out (FIFO). The number of LOFs for one fleet type must equal the number of aircraft of this fleet available, and all LOFs of one fleet type must contain all flights of this type. Because every aircraft of one fleet has to be assigned to the same rotation, all LOFs of each type have to be connected to a circle. Then, if at least every third airport of every LOF is a suitable maintenance station, a valid maintenance routing is found.

The objective of this model is to find a valid routing. Different solutions to the maintenance routing problem are considered to have the same solution quality, thus,

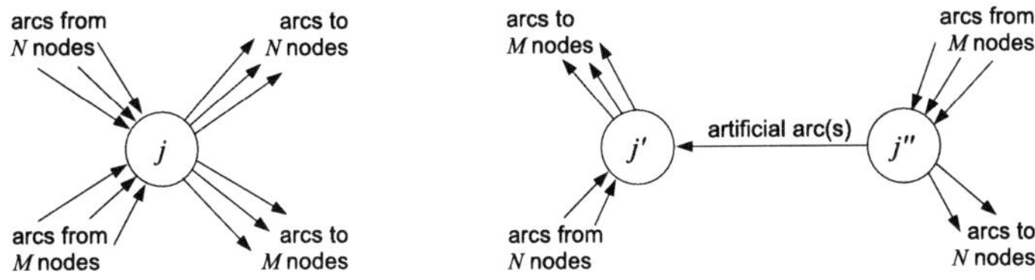

Fig. 4.19 Splitting of nodes and distribution of arcs (Source: Gopalan and Talluri (1998a))

there is no optimization. Given the LOFs, the search for a valid solution represents a network flow problem. A directed graph $G = (V, E)$ represents the stations with aircraft staying at night (vertices V) and the LOFs (arcs E). The number of arcs equals the number of aircraft of that fleet type, one aircraft is assigned to each LOF. V is partitioned into a set M of nodes representing maintenance stations and in a set $N = V \setminus M$ representing non-maintenance stations. In this case, a valid maintenance routing is an Euler tour that includes no more than two nodes of N in succession. This Euler tour should be noted as *three-day maintenance Euler tour (3-MET)*. Assuming that the LOFs are connected and that the number of aircraft of one fleet type departing from one airport is equal to the number of arrival airports (flow balance constraint of the fleet assignment model), G is Eulerian. Gopalan and Talluri (1998a) show that for all stations $j \in N$ the number of arcs m^o_{jM} going out to maintenance stations has to be greater than or equal to the number of arcs m^i_{jN} coming in from non-maintenance stations to provide the existence of a 3-MET. Assuming that this is true for G, a 3-MET in G can be found (if one exists) by searching for an Euler tour in a graph G' derived from G. G is transformed into G' by splitting each node $j \in N$ into two nodes j' and j'' and distributing the incoming and outgoing arcs of j between j' and j'' as shown in Fig. 4.19. A number of $(m^o_{jM} - m^i_{jN})$ additional artificial arcs from j' to j'' are included. Gopalan and Talluri (1998a) show that the existence of an Euler tour in G' implies the extistence of a 3-MET in G. Thus, finding the Euler tour in G' would result in the solution of the maintenance routing problem. The Euler tour can be found by a standard procedure (Bondy & Murty, 1978).

The described procedure represents a polynomial-time algorithm for finding a 3-MET. The 3-MET can only be found if the rotation is connected and the set of LOFs contains a 3-MET ($m^o_{jM} \geq m^i_{jN}$). If the rotation is not connected or if $m^o_{jM} < m^i_{jN}$ two heuristic methods presented by Gopalan and Talluri (1998a) must be applied to transform the LOFs (resp. G) to build a solvable problem. The *Unlocker* tries to construct connected LOFs, whereas the *M-N Improver* modifies the LOFs to meet the second condition. The use of heuristic methods to satisfy the maintenance requirements is necessary because this problem is NP-hard (Gopalan & Talluri, 1998a).

Unlocker. If the LOFs of one fleet type are not connected, the situation is called a locked rotation. Each cycle of connected LOFs is denoted as component. In Fig. 4.20 an example of a locked (two components) and an unlocked (one component) rotation is given. The nodes represent the stations where the aircraft are

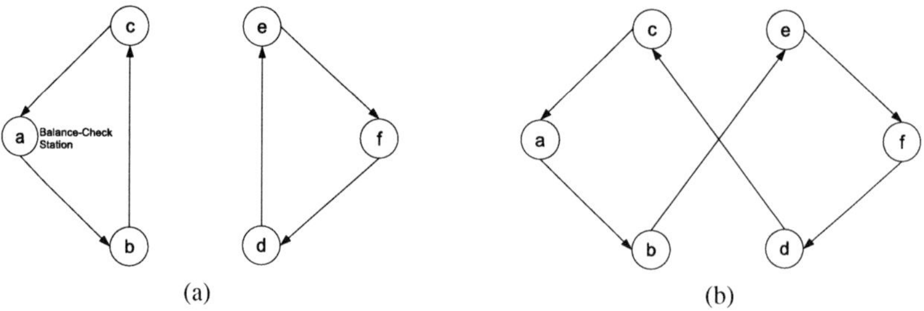

Fig. 4.20 Illustration of a locked (a) and an unlocked (b) rotation (Source: Gopalan and Talluri (1998a))

overnighting, the arcs represent the LOFs. In this example, a and f should represent the maintenance stations with a representing the balance check station, too. There are six aircraft and every aircraft can undergo the maintenance after three days. However, because the graph is locked in figure (a), the aircraft flying between d, e and f will never visit the balance check station a making the routing invalid. In figure (b) the rotation is unlocked fulfilling the maintenance requirements. A graph can be unlocked by changing the flights within the LOFs, for example by swapping the tail assignment between two flights that depart from the same airport at the same time. For example, if the aircraft flying the LOF from b to c and the aircraft flying from d to e are on the ground at some station at the same time (after the turn time has elapsed), their assignment can be swapped leading to the unlocked rotation. In the following, three swapping methods are presented to convert a locked rotation into an unlocked rotation.[12]

1. In the first type of swap, tail numbers are switched between different LOFs within each fleet type, thus, the fleet assignment itself remains unchanged. If multiple components exist for one aircraft type, LOFs (one from each component) have to be found that intersect at one airport at the same time. Then, the flights of the LOFs following this intersection are changed between the LOFs, resulting in modified LOFs unlocking the graph.
2. The first swap mechanism might not be able to unlock the graph because it might be difficult to find an intersection possibility. Thus, the second mechanism changes the fleet type of some flights to unlock the graph. With this swap, only flights within an LOF might be changed, leaving the fleet composition at the end of each LOF unchanged. By changing only these flights, the fleet composition at overnight stations remains unchanged. In addition, changes are rather small, because for each swap the LOFs directly affected are changed; changes are not carried on into connected LOFs via the overnight stations.
3. In the third swapping type the equipment type composition of overnighting aircraft might be changed, thus, increasing the total number of affected flights by the fleet assignment change.

---

[12] Examples for the swapping methods can be found in Gopalan and Talluri (1998a).

If a rotation is locked, the three swap mechanisms are applied in the order presented. Swaps are only allowed if they do not result in new locked rotations of the affected fleets. Each swap has to be feasible. Especially when changing the fleet assignment, different block times of different fleet and operational limitations have to be taken into account. The objective of all three swap mechanisms is to unlock the graph rather than to increase profit. Thus, unlocking the graph might result in less profit because for example a small fleet type needed to be assigned to a high demand flight.

M-N Improver. If in a connected LOF-graph each node in N has fewer arcs coming in from N nodes than arcs going out to M nodes, a 3-MET does not exist, because there are at least three LOFs in succession that do not include maintenance stations at the arriving nodes. The objective of the M-N improver is to fulfill this requirement for each node in N by swapping pairs of edges. Assuming $m^o_{jM} < m^i_{jN}$ for node $j \in N$ and that edge e comes into j from $j_1 \in N$. Edge e will be swapped with an edge e' originating at a node in M. If the terminating node k of e' is a N node, it must satisfy $m^o_{kM} - m^j_{kN} \geq 1$ to fulfill the M-N constraint in k. Assuming that e is an edge going out to a node $j_2 \in N$, a swap of e and e' is allowed if there is a path to all other nodes from the origin nodes of e and e', because then the swap does not create a locked rotation.

Application. The model was implemented in the integrated planning approach as presented in the previous section. For the construction of the LOFs the FIFO-rule was used. However, several enhancements were included to better fit this model into the overall planning process and to increase the chance of finding a feasible solution. For example, the model assumes that the number of LOFs corresponds to the number of available aircraft. However, in this approach changes to the schedule are made in the fleet assignment and optimization step, possibly leading to different numbers of LOFs and aircraft. If aircraft of a fleet type remain unassigned after the construction of the LOFs, new flights for this fleet type are created and included into the schedule. These flights are included in those markets that have the highest market size after realizing all currently scheduled flights (including flights of competing airlines). If these LOFs are not connected (more than one component), the flights are chosen to connect the different components. If the number of LOFs exceeds the number of aircraft of a fleet type, flights need to be removed from the schedule. This task is accomplished by using the operator *Flight Choice* (presented on page 114) that assigns the attribute *optional* to the flights (see page 103) and by restarting the fleet assignment. All three swap mechanisms presented by Gopalan and Talluri (1998a) can only unlock a situation with multiple components of the same fleet type if this is possible with the current flights, thus, the *Unlocker* does not modify the flights except for the assigned fleet. Because this limits the probability of finding a feasible solution, additional unlocking steps are included (*Extended Unlocker*) that are applied when the *Unlocker* presented by Gopalan and Talluri (1998a) fails. The objective is to increase the number of potential positions that allow the swap of aircraft between different LOFs to connect them to one component (two aircraft of two different components have to be on the ground at the same airport at the same time). This is accomplished by including new flights into the schedule. First, additional flights are inserted to connect two components. If additional flights cannot be included because there is not enough time left between the existing flights, the available time is increased by deleting other flights. The number of flights deleted is increased until sufficient connecting flights can be included (at maximum, all flights of one LOF are deleted to allow the unlocking). Like the *Unlocker* of Gopalan and Talluri (1998a), any modification is only allowed if all constraints are satisfied (curfew restrictions, airport operating hours, turn times, operational restrictions etc.). Although this procedure might result in large changes to the schedule with reduced profit, it has to be included to obtain feasible solutions.

The maintenance routing model assumes that each LOF begins with a flight departing after 2 a.m. The routing for each fleet type then is constructed by connecting these LOFs via the overnighting stations to a single circle. However, because the maintenance routing algorithm only considers the origin and destination airport of each LOF when connecting, it might be possible that an LOF starting at 2 a.m. is attached to an LOF ending after 2 a.m., if for example the last flight of the first LOF departs before and arrives after 2 a.m. Although not violating the constraints of the maintenance routing algorithm, this situation would lead to an infeasible routing sequence. In such cases, the routings have to be modified. If there is an LOF that exceeds the departure time of the following LOF, the amount of time that has to be saved to produce a feasible solution is removed from idle ground times between the flights of the LOF. If there is not enough ground time available, two succeeding flights (with the smallest market size) are replaced by one direct flight. To produce a feasible solution, any constraints (curfew restrictions, airport operating hours, fleet ranges, etc.) are taken into account when applying the changes.

4.3.2.3 Schedule Optimization

After the maintenance routing, a feasible airline schedule is given. The objective of the optimization step is to improve this schedule in order to increase the operating profit. In practice, this step is usually performed by human experts supported by DSS to help evaluate their decisions and check for feasibility. Because human experts rely on their experience of many years and often decide by intuition when trying to optimize a given schedule, their procedures are difficult to imitate in an automated solution approach. As a trade-off between this limitation and the potential of computational power, the optimization step in this approach is conducted very extensively, utilizing computational performance to do many incremental search steps to increase a schedule's profit. The development of this optimization step is motivated by the fact that in literature no model could be identified that can be used for schedule optimization and that fits into the gap of the integrated approach presented here.

The optimization is conducted in five steps as presented in Fig. 4.21. The first three steps (*Slack Reduction*, *Airport Removal*, and *Airport Optimization*) are the main optimization algorithms, each representing an iterative greedy improvement heuristic. In each iteration the current schedule is modified by either changing,

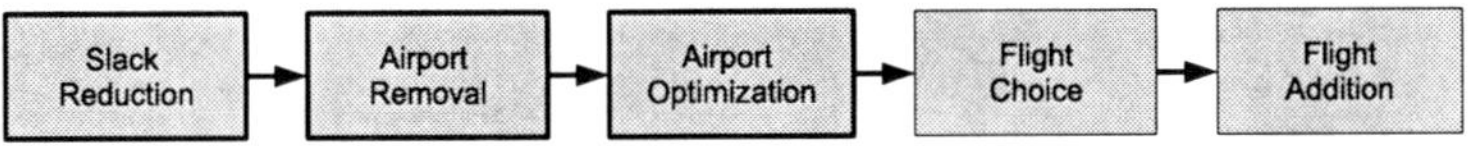

Fig. 4.21 Schedule optimization steps

deleting or inserting flights while maintaining feasibility. A modification is kept if it increases the operating profit. If there is no further increase in profit or a maximum number of iterations $i_{max}^{optimize}$ is achieved, the algorithm stops. $i_{max}^{optimize}$ is determined by a parameter $p_{optimize}$ and the number of flights $|F|$ in the current schedule:

$$i_{max}^{optimize} = p_{optimize} \cdot |F|. \tag{4.51}$$

The last two steps (*Flight Choice* and *Flight Addition*) do not directly change the schedule but work in conjunction with the fleet assignment step in the next iteration of the complete planning approach. Compared to the first three optimization steps, these methods are of minor importance to the overall schedule optimization. In the following, the first three optimization steps are presented including their details as pseudo-code. Then, the last two optimization steps are described.

Slack Reduction. The objective of the step *Slack Reduction* is to minimize idle ground time of aircraft by including new flights into the schedule. Every routing of every aircraft is checked for sufficient ground time between any two succeeding flights. If there is enough time available (the second flight might be shifted backwards), two round-trip flights are inserted at this position. All restrictions have to remain fulfilled. The departure times of both new flights are evenly distributed over the free time period, leading to ground times of the same lenghts between the affected flights. The departure and arrival airport of the round-trip is given by the surrounding flights. Candidates of the airports connecting both flights of the new round-trip are selected by testing all possible airports. All possible round-trip candidates are evaluated according to their contribution to the overall operating profit, the one with the highest contribution is selected. The calculation of the profit contribution of a round-trip candidate requires the evaluation of the schedule including this round-trip, because the connectivity of the flights is changed resulting in different passenger flows. The complete specification of the *Slack Reduction* step is presented as a pseudo-code in algorithm 4.

Airport Removal. The algorithm *Airport Removal* removes the connecting airport of two succeeding flights and, thus, replaces the two flights by one direct flight. This procedure allows the reduction of the number of less profitable flights and creates more slack time that could be filled with new flights using the *Slack Reduction* procedure. Given all candidates of airports between two connecting flights, the one resulting in a maximum increase in operating profit is removed. The complete specification of the *Slack Reduction* step is presented as a pseudo-code in algorithm 5.

Algorithm 4. Slack Reduction

1: read schedule F
2: read available set of airports A
3: $i = 0$
4: **repeat**
5: 　$i = i + 1$
6: 　$F_m = F$　　　　　　{F_m is working schedule}
7: 　$F_t = F$　　　　　　{F_t is temporary best schedule}
8: 　**for all** flights f of F **do**
9: 　　e is direct predecessor of f
10: 　　g is direct successor of f
11: 　　calculate available ground time t_{ef} between e and f
12: 　　calculate available ground time t_{fg} between f and g
13: 　　**for all** airports a of A **do**
14: 　　　create flights x and y connecting e and f via a
15: 　　　**if** x, y satisfy fleet restrictions
　　　　and x, y satisfy maintenance restrictions
　　　　and x, y satisfy curfew restrictions **then**
16: 　　　　**if** $t_{ef} + t_{fg}$ long enough for x and y **then**
17: 　　　　　$f_m = f$
18: 　　　　　**if** t_{ef} not long enough for x and y **then**
19: 　　　　　　postpone f_m until t_{ef} long enough for x and y
20: 　　　　　**end if**
21: 　　　　　**if** f, x, y satisfy airport operating hours **then**
22: 　　　　　　$F_m = F$
23: 　　　　　　replace f by f_m
24: 　　　　　　include x and y between e and f_m in F_m
25: 　　　　　　**if** profit of F_m higher than profit F_t **then**
26: 　　　　　　　$F_t = F_m$
27: 　　　　　　**end if**
28: 　　　　　**end if**
29: 　　　　**end if**
30: 　　　**end if**
31: 　　**end for**
32: 　**end for**
33: 　**if** profit of F_t higher than profit F **then**
34: 　　$F = F_t$
35: 　**end if**
36: **until** profit of F_t lower than profit F
　　　or $i > i_{max}^{optimize}$
37: return schedule F

Algorithm 5. Airport Removal

1: read schedule F

2: read available set of airports A

3: $i = 0$

4: **repeat**

5:    $i = i + 1$

6:    $F_m = F$ $\{F_m$ is working schedule$\}$

7:    $F_t = F$ $\{F_t$ is temporary best schedule$\}$

8:    **for all** flights f of F **do**

9:       e is direct predecessor of f

10:      g is direct successor of f

11:      h is direct successor of g

12:      create flight x connecting e and h

13:      **if** x satisfies fleet restrictions

          and x satisfies maintenance restrictions

          and x satisfies curfew restrictions

          and x satisfies airport operating hours **then**

14:        $F_m = F$

15:        include x, remove f and g from schedule F_m

16:        **if** profit of F_m higher than profit F_t **then**

17:          $F_t = F_m$

18:        **end if**

19:      **end if**

20:    **end for**

21:    **if** profit of F_t higher than profit F **then**

22:      $F = F_t$

23:    **end if**

24: **until** profit of F_t lower than profit F

     or $i > i_{max}^{optimize}$

25: return schedule F

Airport Optimization. The *Airport Optimization* step changes airports of connecting flights to improve the schedule. Each airport represents a departure airport and arrival airport of two flights, thus, if the airport is changed, the two corresponding flights are replaced. The new airport and the one to be replaced are selected according to the highest increase of the operating profit by the two corresponding new flights. The complete specification of the *Airport Optimization* step is presented as a pseudo-code in algorithm 6.

Algorithm 6. Airport Optimization

1: read schedule F

2: read available set of airports A

3: $i = 0$

4: **repeat**

5:      $i = i + 1$

6:      $F_m = F$ {F_m is working schedule}

7:      $F_t = F$ {F_t is temporary best schedule}

8:      **for all** flights f of F **do**

9:          e is direct predecessor of f

10:         g is direct successor of f

11:         h is direct successor of g

12:         calculate available time t_{eh} between e and h

13:         **for all** airports a of A **do**

14:             create flights x and y connecting e and h via a

15:             **if** x,y satisfy fleet restrictions

                    and x,y satisfy maintenance restrictions

                    and x,y satisfy curfew restrictions

                    and x,y satisfy airport operating hours

                    and t_{eh} long enough for x and y **then**

16:                 $F_m = F$

17:                 replace f and g by x and y in F_m

18:             **end if**

19:         **end for**

20:     **end for**

21:     **if** profit of F_t higher than profit F **then**

22:         $F = F_t$

23:     **end if**

24: **until** profit of F_t lower than profit F

        or $i > i_{max}^{optimize}$

25: return schedule F

Flight Choice. The objective of this step is to remove non-beneficial flights from the flight schedule. The number of flights being subject to removal is controlled as percentage $p_{opt} \in [0, 1]$ of the total number of flights in the schedule. Then, the flights with the lowest profit contribution are chosen as candidates for deletion. Instead of immediately removing the flights, this operator assigns the attribute *optional* to these flights. The fleet assignment actually removes the flights if they are not necessary to meet fleet assignment constraints.

Flight Addition. In this step, new round-trip flights are created and included in the schedule. The number of new flights is chosen according to a percentage $p_{new} \in [0,1]$ of the number of flights in the current schedule. The markets in which new flights are added are chosen randomly according to the size of the remaining demand in this market (high demand markets receive a higher selection probability). The parameter $tw_{new} = 2 \cdot tw$ controls the length of the time window for the fleet assignment step (see page 103). The larger time window increases the flexibility which is necessary to fit the new flights into the schedule. The probability of assigning the attribute *optional* for the new flights is $1 - p_{opt}$.

4.3.3 Solution Process

4.3.3.1 Supportive Functions

The planning and optimization steps presented in the previous section represent the main tasks necessary for airline scheduling. Every task requires specific input data to produce a feasible (or optimal) solution. Because not always every preceding planning step fulfills these constraints, additional effort is necessary to support the linkage between the single steps and to find feasible or optimal solutions in each step. For example, the fleet assignment model requires flow balance at every airport for every fleet. If this constraint is not satisfied by the flight schedule given as input, the fleet assignment cannot be solved, and the flight schedule has to be modified. Another example is the maintenance routing: if there is no flight to a maintenance station, new flights must be included or existing flights must be modified to include overnighting aircraft at maintenance stations. In practice, manual inputs and human interaction would be the linkage between the individual solution steps and experienced experts would change given input if it resulted in infeasible subsolutions. In the following, supportive functions are presented that are used in this sequential planning approach between the solution steps if a subproblem cannot be solved.

Balance Aircraft Flow. To solve the fleet assignment problem, flow balance is required: for each fleet type and airport, the number of aircraft arriving has to equal the number of departing aircraft. To prevent flow imbalance during schedule construction, the *Balance Aircraft Flow* algorithm modifies the flights of any given schedule to meet this constraint. Because every flight has a departure and arrival airport, imbalances always affect at least two airports. If one airport has more arriving flights than departing flights, there is at least one airport with the reverse situation. The algorithm tries to fix this problem by replacing the flight causing the imbalance with a new one. If this cannot be performed (for example because the flight would be infeasible for the given aircraft type or curfew restrictions would be violated), additional flights are changed until the flow is balanced at all airports. The probability of selecting a flight for modification follows the market size: the smaller the market size of an unbalanced flight, the higher the chance that this flight is changed or removed.

Flight Choice. The fleet assignment is infeasible if too many flights exist that have to be assigned to all aircraft or aircraft of one fleet type. The *Flight Choice* step changes the attribute *optional* of the flights indicating whether the fleet assignment model might remove the flight from the schedule. This step corresponds to the *Flight Choice* step presented on page 112.

Increase Connectivity. One goal of the maintenance routing is to construct one rotation per fleet. Thus, it is necessary that all flights can be ordered in a sequence that can be flown by one aircraft. In addition, the fleet assignment problem might be difficult to solve or even infeasible if there are a lot of flights spread over many different airports. In this case, meeting the flow balance could be difficult to achieve for each fleet type. The *Increase Connectivity* step assists in both the following cases, maintenance routing and fleet assignment. This algorithm changes, removes and inserts flights that increase the connectivity within the airline's flight network. The number of flights being subject to modification is controlled as percentage $p_{cnx} \in [0, 1]$ of the total number of flights. The selection of flights to be changed follows the amount of traffic at the airports: the more traffic at one airport according to the actual flights, the higher the chance that this airport will get additional (departing and arriving) flights, and vice versa. Any imbalances at the airports additionally increase the probability of modifying related flights to reduce the imbalance. In addition, the *Flight Choice* step is conducted, also simplifying the fleet assignment and the maintenance routing, because there are fewer flights that need to be assigned to a fleet or rotation.

Insert Maintenance Flights. The maintenance routing problem can only be solved if there are enough flights to a maintenance station at the end of each day. If there is no flight to a maintenance station, the three-day maintenance routing is infeasible and even the advanced mechanisms presented in Sect. 4.3.2.2 cannot produce a valid routing. If there are no sufficient flights to or from the maintenance stations for each fleet type, a flight departing from a maintenance station is created and included into the schedule for each affected fleet type. The maintenance station and the arrival airport are chosen randomly, the departure time is as early as possible. To prevent flow imbalances, the *Balance Schedule* mechanism needs to follow each insertion of a new maintenance flight. The new flight is necessary to comply with the maintenance constraints but might have a poor profit share. To prevent the removal of this flight by the fleet assignment or assignment of another fleet to minimize costs, this flight receives the attribute *maintenance* (see page 103) indicating that the flight is fixed and may not be changed or removed by the fleet assignment algorithm. However, the optimization steps might change this flight because these steps always consider the three-day-maintenance routing constraint when applying changes to the schedule. For example, a new maintenance flight with a poor profit share might be modified regarding its destination airport or departure time. In addition, the departing maintenance station might be exchanged with another maintenance station for this flight.

Use Optimized Schedule. The sequential approach represents an iterative procedure: after the optimization steps presented in Sect. 4.3.2.3 the complete process starts again with the flight scheduling and fleet assignment step. Then, the fleet assignment might be infeasible due to the new additional flights that were inserted with the last optimization step *Flight Addition*. The *Use Optimized Schedule* then assigns the attribute *optional* (see page 103) to some new flights and removes this attribute from old flights (that was set by the *Flight Choice* step), relaxing the fleet assignment problem. More than one attempt might be necessary until the fleet assignment problem can be solved, in each attempt the number of affected flights is increased. If this percentage reaches 100%, the *Flight Choice* and *Flight Addition* steps are revoked and the schedule produced by the *Airport Optimization* representing the last feasible solution step is processed by the fleet assignment.

4.3.3.2 Schedule Initialization

By combining the solution steps and the supportive functions to the overall planning procedure, an airline schedule can be optimized following the sequential and iterative planning paradigm. However, the presented mechanisms need an initial schedule to start with. Because the maintenance routing algorithm and the optimizing steps *Slack Reduction* and *Flight Addition* iteratively insert new flights into the schedule, an initialization method only needs to construct a very basic schedule which will then be improved and extended using the procedures mentioned. For this approach, this basic initial schedule is created by constructing one flight for each aircraft available. These flights are created in those markets with the highest remaining market size (passenger demand after subtracting passengers currently traveling on the competitors' flights and own flights already included in the schedule). Departure times are the peak times of the demand distribution over the day (see Fig. 4.12). To be solvable by the first fleet assignment step, the *Balance Aircraft Flow* function is applied. Then, after the fleet assignment, the maintenance routing is likely to insert additional flights because the number of LOFs should be smaller than the number of aircraft and the LOFs might not be connected. After maintenance routing, the *Slack Reduction* method will insert additional flights, because there is a lot of idle ground time available since there is only one flight per aircraft in the schedule.

To summarize, schedule initialization does not consist of one single function creating an extensive and acceptable first schedule; instead, it provides a very basic schedule that then is iteratively extended with new flights and improved using the solution steps and supportive functions.

4.3.3.3 Integration

The main challenge in constructing a complete airline scheduling procedure based on the described methods is to link the single steps so that each can solve its subproblem based on the given output of the preceding step and produce a feasible solution. In addition, since the procedure should be able to construct schedules for any given setting, there should be no restrictions on the given input data. After an initial schedule is created, the fleet assignment, maintenance routing, and schedule optimization are conducted; then, these three steps are performed iteratively until the optimizing steps cannot improve the schedule any more or until there was no increase in profit after i_{max} iterations.

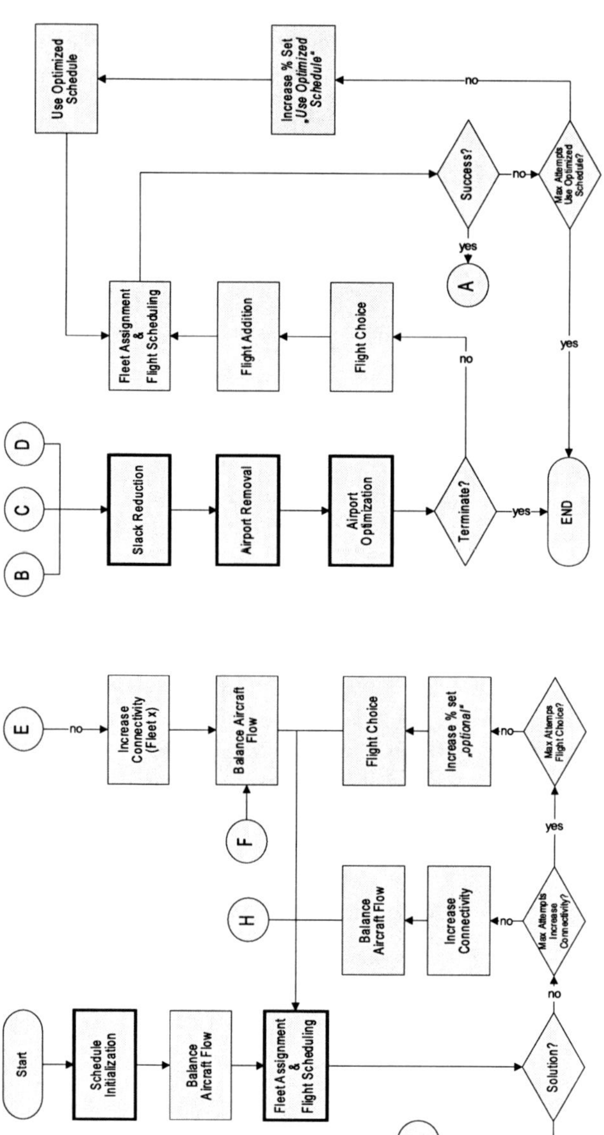

Fig. 4.22 Sequential airline scheduling approach flowchart (part 1)

In the following figures, an overview of the complete sequential approach is presented as a flowchart. Because of its complexity, the total procedure is split up into two figures 4.22 and 4.23. Transitions between both figures are denoted as circles with appropriate letters. The main solution and optimization steps can be identified by the boxes with thick frames.

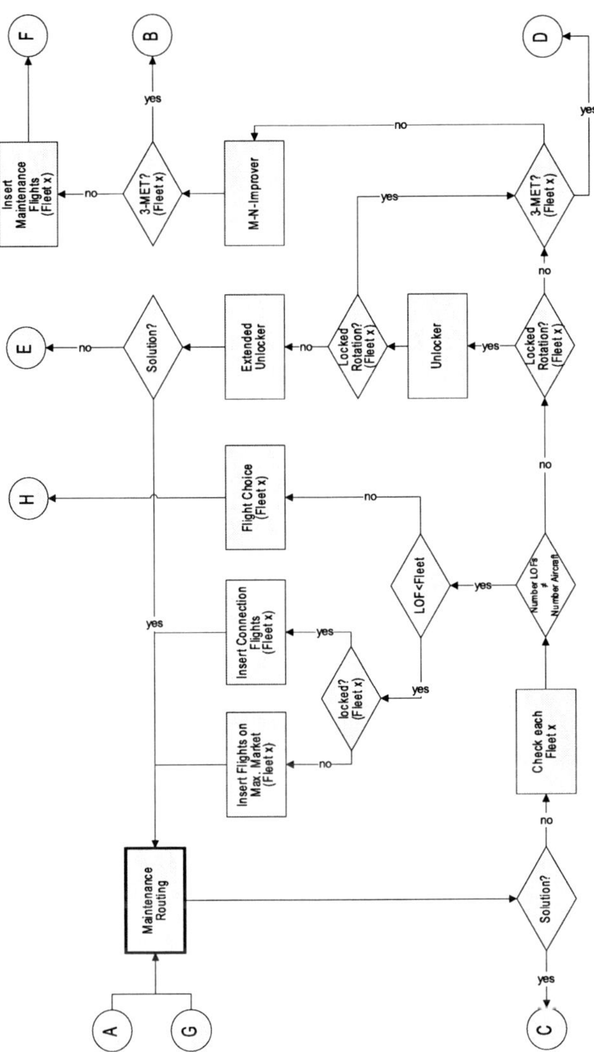

Fig. 4.23 Sequential airline scheduling approach flowchart (part 2)

4.3.4 Experiments

In this section, results from implementing and applying the complete sequential airline scheduling approach are presented. First, calibration results focusing on the setting of the models parameters are presented in Sect. 4.3.4.1. Then, in Sect. 4.3.4.2, the obtained solutions and the search process of the calibrated model are analyzed.

The sequential planning approach has been implemented in the C++ programming language. One ILOG CPLEX 9.0 license was available for the fleet assignment step, standard parameters were used. To allow an efficient use of this license, a distributed implementation was developed that calculates the fleet assignment and routing on the workstation with CPLEX and conducts the computationally intensive optimization steps on different workstations with different processor and memory specifications.

4.3.4.1 Calibration

The presented planning approach consists of individual solution steps and supportive steps that are controlled by parameters. The objective of the calibration process presented in the following is to find a parameter setting that yields the best results of the solution approach. The complete set of parameters that has to be calibrated are as follows:

- number of iterations i_{max} of the complete procedure without increase in profit to determine the termination of the optimization (see page 116),
- time window size tw in minutes of the current flights in the combined flight scheduling and fleet assignment step (see page 103),
- percentage p_{cnx} to determine the number of flights to be modified by the *Increase Connectivity* step (see page 114),
- percentage p_{opt} to determine the number of flights set to the attribute *optional* by the *Flight Choice* step (see page 112) ,
- percentage p_{new} to determine the number of new flights inserted by the *Flight Addition* step (see page 113),
- percentage $p_{optimize}$ to determine the maximum number $i_{max}^{optimize}$ of attempts conducted by each optimization step *Slack Reduction*, *Airport Removal*, and *Airport Optimization* (see page 109).

Different parameter settings are examined by applying the planning approach to five different planning scenarios (see Appendix B), representing an airline's possible starting point when facing the airline scheduling problem. The values for the parameters are calibrated separately (*ceteris paribus*): one parameter is set to different values while the other parameters remain constant. The values of the constant parameters are chosen according to the following:

$$i_{max} = 5,$$
$$tw = 60,$$
$$p_{cnx} = 0.3,$$
$$p_{opt} = 0.3,$$
$$p_{new} = 0.1,$$
$$p_{optimize} = 0.1.$$

The results of the different parameter settings focus on two key figures:

1. the objective value, and
2. the number of fitness evaluations until the best solution was found, representing a platform-independent quantification of the effort necessary to find the best solution.

For each scenario and setting, five optimization runs are conducted. The results of the different planning scenarios vary in their order of magnitude, because the scenarios consist of different numbers of aircraft and airports. Thus, to find a parameter setting based on all scenarios, normalization and aggregation of the results are necessary. In this study, for a given parameter setting and scenario the deviation of the averaged results from the mean value of all parameter settings is used as an indication of the current setting's quality. Let $f_{p,s}$ denote the average fitness value of the five runs with parameter setting $p \in P$ (P is the set of tested values for p) and scenario $s \in S$, then the average fitness value $\bar{f}_s$ for all settings for scenario s is calculated as:

$$\bar{f}_s = \frac{\sum\limits_{p \in P} f_{p,s}}{|P|}.$$ (4.52)

The impact $i_{p,s}$ of setting p in scenario s is expressed as a relation with this average fitness:

$$i_{p,s} = \frac{f_{p,s} - \bar{f}_s}{|\bar{f}_s|}.$$ (4.53)

Finally, the aggregation of all scenarios yields the setting p's average impact $\bar{i}_p$ on solution quality:

$$\bar{i}_p = \frac{i_{p,s}}{\sum\limits_{s \in S} i_{p,s}}.$$ (4.54)

The following figures present the results $\bar{i}_p$ for different settings p for each parameter as smoothed curves. The results of the number of fitness evaluations are calculated accordingly.[13]

Fig. 4.24 presents the calibration of the parameter i_{max}. As was expected, an increase of the solution quality can be observed for higher values of i_{max}. A higher i_{max} gives the solution approach more time to find a better solution and to escape from local optima. However, if i_{max} is further increased, surprisingly solution quality

---

[13] The individual results of the scenarios including the absolute values are presented in Sect. C.1.1 in the appendix.

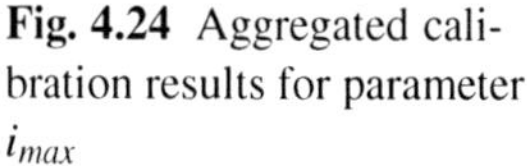

Fig. 4.24 Aggregated calibration results for parameter i_{max}

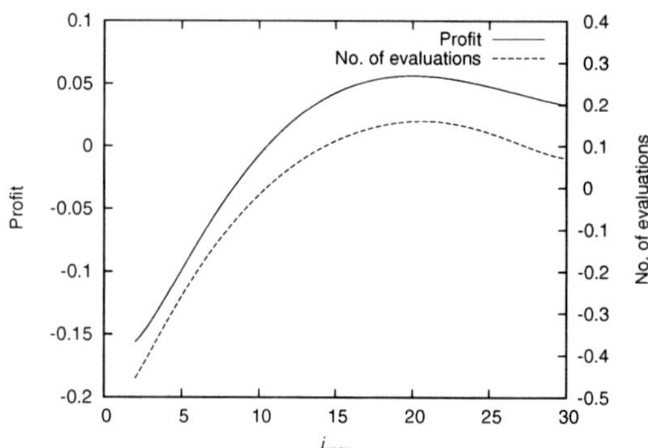

decreases. In future work, this observation has to be further investigated including additional experiments for confirmation.

Fig. 4.25 presents the calibration of the time window size *tw*. Increasing the time window size leads to higher computation times, since there is more freedom in planning. Solution quality reaches its maximum at approximately 30 minutes, further increasing this parameter results in lower solution quality. An explanation for this effect might be the assumption within the fleet assignment step of uniformly distributed demand within each time window. If this time window is long, large variations in the departure times will have an effect on passenger demand that is not detected by the fleet assignment step.

Fig. 4.25 Aggregated calibration results for parameter *tw*

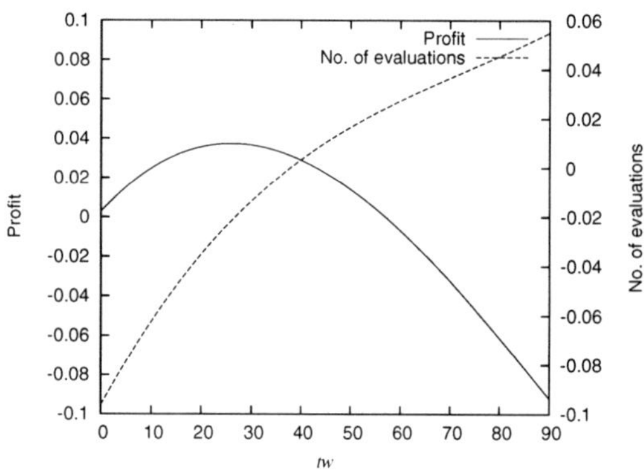

Fig. 4.24 presents the calibration of the parameter p_{cnx}. As can be observed, very small and high p_{cnx} results in lower solution quality, best solutions were obtained with $p_{cnx} = 0.05$. If p_{cnx} is close to 0, the *Increase Connectivity* step only applies very small changes to the schedule if it is infeasible. This might not be sufficient to obtain a feasible solution, which then has to be constructed using more rigorous operators that reduce solution quality. In contrast, if p_{cnx} is set to high values, in each step many modifications to a schedule are applied independently of profit considerations and, thus, as a result solution quality is limited.

Fig. 4.26 Aggregated calibration results for parameter p_{cnx}

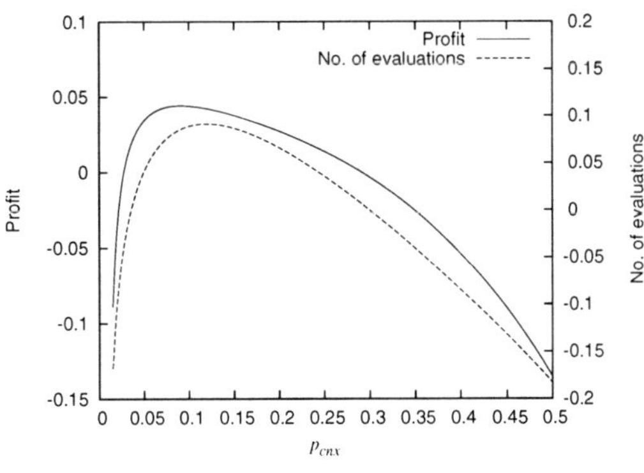

Fig. 4.27 Aggregated calibration results for parameter p_{opt}

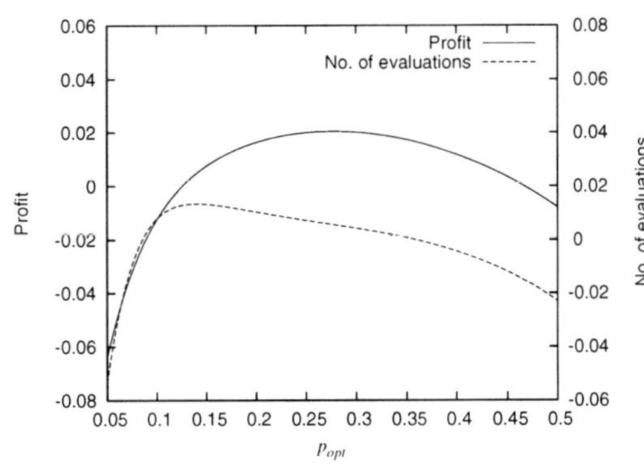

Fig. 4.28 Aggregated calibration results for parameter p_{new}

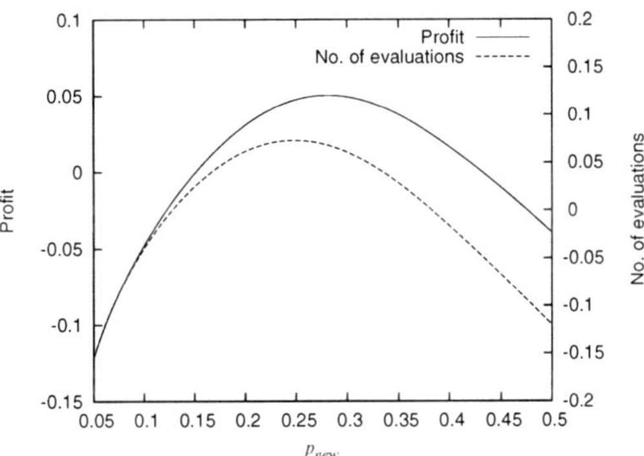

Figures 4.27 and 4.28 present the calibration of the parameters p_{opt} and p_{new}. In general, the results are very similar. Parameter values around 0.25 yielded best results. The closer each parameter is to 0, the less the effect of the function using this parameter, and vice versa. Thus, the observed values for the best solution quality are presumed to result in the best compromise of a too low and a too high impact of each corresponding technique.

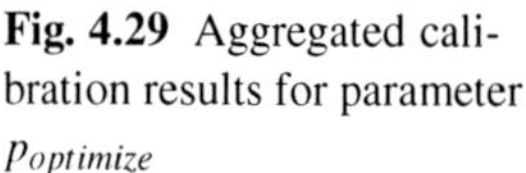

Fig. 4.29 Aggregated calibration results for parameter $P_{optimize}$

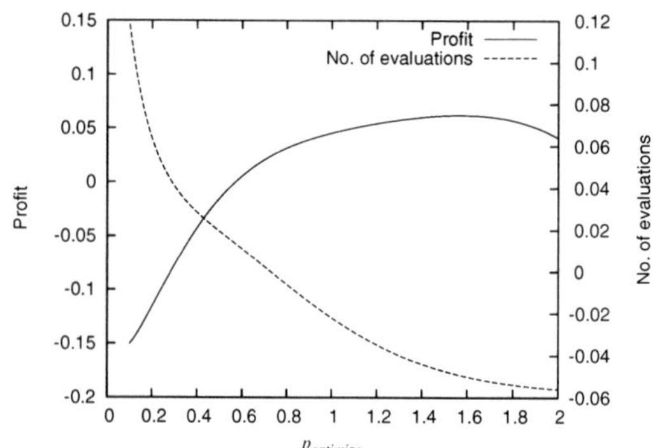

Fig. 4.29 presents the calibration of the parameter $p_{optimize}$ from which the number $i_{max}^{optimize}$ of applications of each optimization method is determined. Increasing $p_{optimize}$ leads to more optimization steps in each iteration, thus, solution quality increases. The decreasing computation time might be explained by the reduced influence of the other steps on the overall solution approach. The higher the number of optimization steps in one iteration, the more the overall schedule is determined. The degrees of freedom decrease (for example, there is less slack time), thus, there is less room for other solution steps to guide the solution towards their objectives that are not congruent with the operating profit and would increase computation time.

For each parameter, using the value at which the best fitness was achieved results in the following final parameter setting, that is used for the subsequent experiments and analyses:

$$i_{max} = 20,$$
$$tw = 30,$$
$$p_{cnx} = 0.05,$$
$$p_{opt} = 0.25,$$
$$p_{new} = 0.25,$$
$$P_{optimize} = 1.5.$$

4.3.4.2 Analysis

In the following, the solutions obtained and the solution process are analyzed. For this purpose, the calibrated planning approach is applied to the five planning scenarios.

To analyze the solutions obtained by the sequential planning approach, the following Table 4.7 presents some (average) key figures of the schedules (standard deviations in parentheses). Because of the high amount of information in each schedule, presenting each schedule of every optimization run for each planning scenario would be beyond the scope of this study. In addition, airline schedules

Table 4.7 Key figures of airline schedules constructed with the sequential planning approach

Key Figure	scenario				
	A	B	C	D	E
Profit	450,629	325,927	-60,166	51,849	97,774
	(59,668)	(16,536)	(20,219)	(20,835)	(7,345)
SLF	0.300	0.369	0.172	0.387	0.267
	(0.012)	(0.021)	(0.017)	(0.023)	(0.027)
No. of passengers	5,947	3,493	1,682	2,144	2,057
	(616.36)	(183.13)	(65.06)	(48.15)	(270.72)
No. of flights	136	97	61	73	49
	(10.55)	(8.50)	(3.19)	(3.97)	(4.90)
No. of fitness evaluations	212,886	36,772	154,618	32,265	48,621
	(183,070)	(11,386)	(85,735)	(12,894)	(35,137)
Total no. of evaluations	369,150	112,090	260,428	50,335	83,369
	(236,341)	(25,747)	(137,999)	(10,100)	(29,406)
No. of iterations	19	24	36	9	20
	(12)	(5)	(17)	(3)	(5)

appearing to be different could include very similar flight programs (for example, flights are assigned to a different fleet type or another rotation), limiting the explanation of a very detailed presentation.[14] In general, as the low standard deviations indicate, the results are stable. Most variations exist in the duration of the complete optimization run, which is measured by the last three rows of Table 4.7. Scenario A resulted in the highest profit values, although the seat load factor was best for scenario D. Scenario C even results in an operational loss (although it required the most attempts to improve solution quality, since the number of iterations is highest). Compared to the uncalibrated model (basic parameter setting), the obtained profit values represent an average increase of 19.26%. The number of required fitness evaluations is on average 38.39% higher than with the uncalibrated model. Differences in the order of magnitude between the number of iterations and the number of fitness evaluations are the result of the different specifications of the scenarios: the more aircraft and airports are available in a scenario, the more fitness evaluations are necessary in each iteration. It has to be emphasized that a meaningful interpretation of the absolute values of these indicators is not possible, since for each scenario the competition and the set of airports were chosen randomly (and, thus, can represent markets with low airline travel demand) and the market size estimates represent only poor approximations of the real demand. In reality, airlines usually have average SLF of about 75% ICAO (2006).[15]

The following figures focus on the solution process. They present results of experiments on scenario A as a representative example for all scenarios.[16] Because the

---

[14] Individual results are presented in tables in Sect. C.2.1.1.

[15] For example, when optimizing airline schedules for scenario D but using past passenger numbers for selected city pairs as market sizes instead of the estimates (see the discussion on 75), a SLF of 0.630 was obtained.

[16] The results of all scenarios are presented in Sect. C.2.1.2 in the appendix.

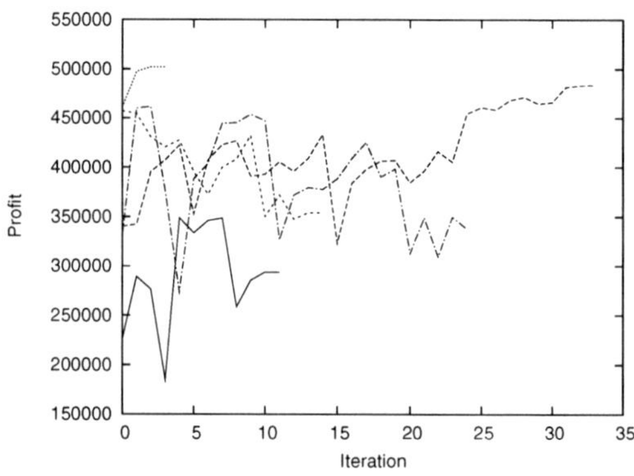

Fig. 4.30 Trend of profits of all five optimization runs of scenario A

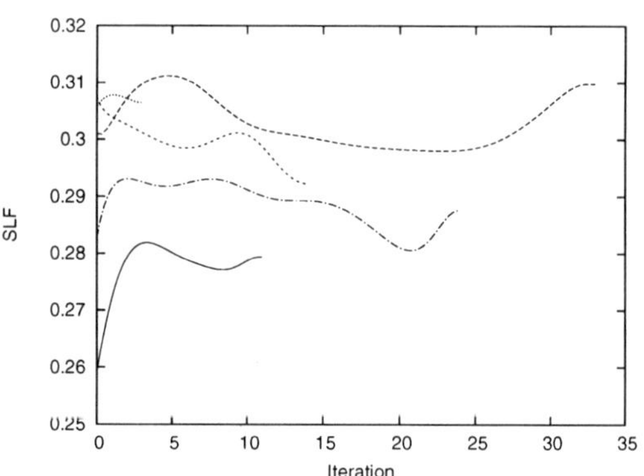

Fig. 4.31 Trend of seat load factors (SLF) of all five optimization runs of scenario A

number of iterations is different between the individual runs for each scenario and between the scenarios, a meaningful aggregation among the individual runs is not possible.

Fig. 4.30 plots the profit for the five different runs of scenario A. As is clearly visible, the number of iterations varies among the different runs, the shortest run required less than five iterations until it terminated. However, it yielded the best solution quality. Besides a general growth from start to end (except for one run), the optimization progress shows a very unstable trend. There are large variations (peaks and drops) in the profit even between succeeding iterations. As will be shown later in Fig. 4.34, this is most likely the result of the maintenance routing steps and the related supportive functions.

The following three figures show similar characteristics. They plot the SLF (Fig. 4.31), the number of flights (Fig. 4.32), and the number of passengers (Fig. 4.33) as smoothed curves. The number of flights and the number of passengers have the same trend. Thus, if there are more flights, more passengers are transported. Except for one run, the number of flights increases in the beginning. This trend results from the basic initialization of the schedule, in which only a basic schedule

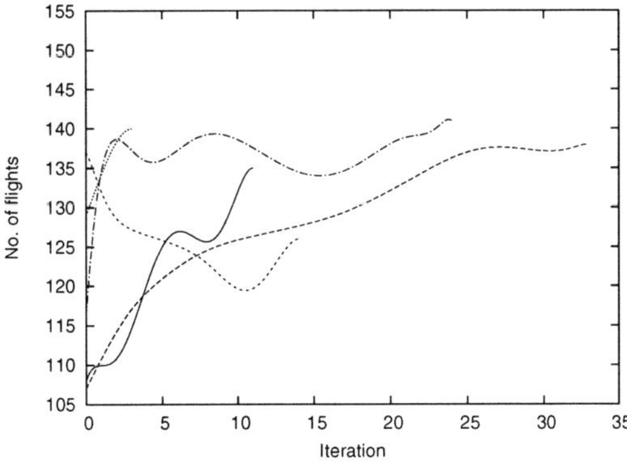

Fig. 4.32 Trend of numbers of flights of all five optimization runs of scenario A

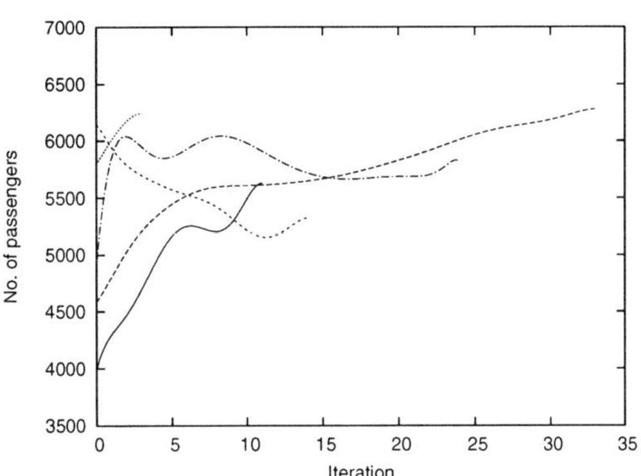

Fig. 4.33 Trend of numbers of passengers of all five optimization runs of scenario A

is created, which is then successively extended by the solution steps during the iterations.

To investigate the contribution of the different solution steps of the sequential planning approach, the following Fig. 4.34 plots the profit on a more detailed level (for each solution step for each iteration). For clarity, the results of only one run of scenario A are presented as a representative example.[17] To better understand the shape of the plot, Fig. 4.35 presents an excerpt of Fig. 4.34, focusing on iterations 3-5. As this figure illustrates, the drops in operating profit result from the application of the maintenance routing algorithm. Thus, the results from the preceding fleet assignment usually contain locked rotations and/or are infeasible with respect to the three-day maintenance requirement. Because the repair mechanisms of the maintenance routing step do not consider the operating profit when modifying the current schedule, the drops can easily be explained.[18] A profit decrease in

---

[17] Similar figures for the other runs and for the other scenarios are presented in Sect. C.2.1.2 in the appendix.

[18] However, the additional repair mechanisms developed in this study take market sizes into account when applying changes to the schedule. For example, if a flight needs to be inserted, this is accomplished for the market with the highest passenger demand (under consideration of demand already satisfied by existing flights).

Fig. 4.34 Profit contribution by individual solution steps (scenario A, run3)

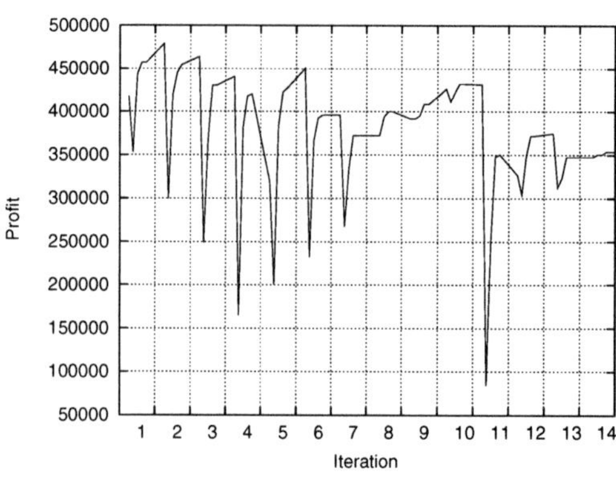

Fig. 4.35 Profit contribution by individual solution steps between iteration 3 and 5 (scenario A, run3)

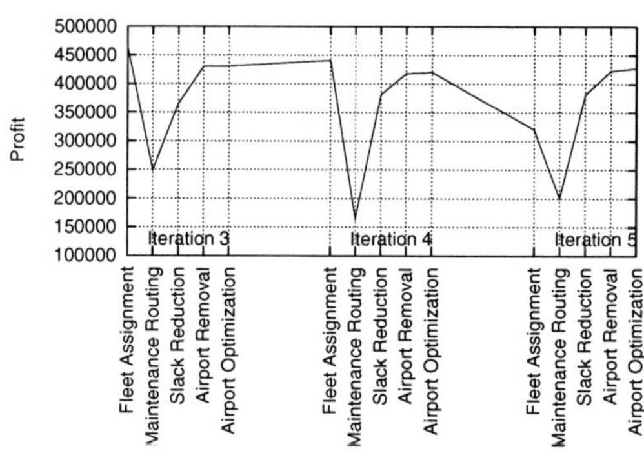

the fleet assignment step (for example in iteration 5 in Fig. 4.35) can be explained by the objective function of this step, which is to minimize costs (operating and spill costs). Because revenues are not taken into account, minimizing operating costs can contradict maximizing profit. For example, if no flights at all are conducted, operating costs are minimized without earning operating profit.

4.3.5 Summary and Conclusion

4.3.5.1 Summary

In this section, an integrated airline scheduling approach was presented that follows the traditional airline scheduling process consisting of separate planning steps solved in a sequence. The procedure presented here has three major planning steps: a combination of flight scheduling and fleet assignment, maintenance routing, and schedule optimization. For the first steps, solution models from OR literature were used. The optimization step consists of three greedy algorithms each using a different local search operator that iteratively modifies a schedule until no further improvement of operating profit can be achieved. All models have special requirements with regard to their input data to produce feasible solutions. In practice, many manual inputs and feedback loops are necessary to apply the models to airline scheduling; in the scheduling procedure presented here, these assisting decisions are made by some modifications to the existing (original) models and by supportive functions. Their objective is to better link the individual solution steps. If a feasible solution cannot be obtained by one model, its input is modified in order to increase the chance of finding a good solution.

Parameters that control the overall planning procedure were obtained by testing various settings with regard to the operating profit for five different planning scenarios. The calibrated model was then applied to these scenarios for an analysis of the obtained solutions and the search process. The solutions of the different runs are stable with regard to the resulting operating profit, number of flights, number of passengers etc. However, differences exist in the effort of the optimization runs. There are large variations in the number of fitness evaluations and in the number of solution iterations. One factor contributing to this observation is the unstable optimization progress which is characterized by peaks and drops of the profit even between succeeding iterations. A closer look at the profit contribution of each solution step within an iteration unveils the fact that the application of the maintenance routing step is most likely responsible for the drops, which are then again compensated by the optimization steps. Since the fleet assignment preceding the maintenance routing might produce schedules that consist of locked rotations or do not fulfill the maintenance requirements, the schedules have to be repaired, which is conducted regardless of the profit.

4.3.5.2   Conclusion

Because of the stepwise approach, supportive functions had to be included that assist each step to find a feasible solution and to integrate the individual solution steps into one iterative procedure, leading to a rather complex planning procedure. The (simplified) flowchart in figures 4.22 and 4.23 gives an impression of this complexity. Consequently, different parameters had to be set to control the planning procedure.[19] Their calibration was conducted ceteris paribus. Because interdependencies between the parameters are likely to exist, an extensive calibration process in which all possible parameter combinations are tested should be conducted in future work. In addition, some alternative options within each solution step could be tested to find out whether they would result in more profitable solutions (for example the LIFO-rule instead of the FIFO-rule in maintenance routing or the number of flight arcs per flight in the fleet assignment model). Another option is to change the order in which the three individual optimization steps are conducted.

A major drawback of the presented approach is its sequential planning paradigm. Each solution step has a different objective function which conflict to some extent. For example, the objective of fleet assignment is to minimize costs (operating and

---

[19] Many more additional, but less important parameters could be selected in some of the planning approach's functions. Since unlimited effort could be made to test all values for all parameters, they were set by common sense.

spill costs), not to maximize revenue. Thus, minimizing operating costs could be realized by conducting only a small number of flights.[20] In addition to conflicting objectives, constraints of one solution step can often not be fulfilled based on the given input from the preceding step. For example, the maintenance routing has to find a routing that is feasible with regard to maintenance restrictions. For this purpose, the flights and the fleet assignment might be changed, reducing the solution quality. Although in this study these modifications take market sizes into account, this effect is rather strong because sometimes it is difficult to find a feasible routing. In such cases, many changes are applied to the schedule, often resulting in a much lower solution quality. This effect can be observed in figures 4.34 and 4.35, in which a decrease in solution quality follows the maintenance routing step.

In addition to some drawbacks resulting from the stepwise planning paradigm and the inadequate linkage between the steps, the individual solution steps inhibit some limitations. For example, the optimization steps are straightforward but not exclusive. They represent local search operators, which could be further improved by enlarging the neighborhood related to each modification or by changing the type of modification (for example, changing the fleet assignment). A second example is the maintenance routing step, representing a simplified model of the real maintenance problem that does not take all practical requirements into account. There is no capacity constraint for maintenance at the airports; in general, there could be a solution with every aircraft undergoing maintenance at the same airport on the same day, which would be unrealistic in practice. In addition, it is assumed that maintenance always takes place at night after at least three days, there is no consideration of the real flight hours conducted and the minimal time of duration required by maintenance. In practice, maintenance is performed after a maximum number of flight hours or landings and requires a specific amount of time, varying across different fleet types and airports.

On the other hand, the separation of the different solution steps allows a straightforward improvement of the individual tasks. Necessary input and output data of each step is known (and – if necessary – modified by the supportive functions), and each individual procedure can be replaced by an improved version. These improvements can also include an extension of the scope and further integration, reducing the amount of supportive functions necessary. For example, if an improved fleet assignment model including maintenance consideration could be implemented, the destructive effects of the maintenance routing algorithm might be reduced possibly leading to a much more progressive optimization process.[21] Since many different procedures are included in the overall planning approach (solutions steps and supportive functions), there are many starting points for further improvements and enhancements.

---

[20] This applies to flights with the attribute *optional*, since all other flights have to be conducted

[21] Of course, an extended model must include at least the capability of the previous model. In this example, an improved fleet assignment model still has to decide on the flight scheduling.

4.4 Simultaneous Approach

4.4.1 Overview

In this section, a simultaneous approach for airline scheduling is presented. Like the sequential approach, the flight schedule generation and aircraft scheduling phases from Fig. 2.2 are included. The objective of this approach is to overcome the artificial decomposition of the overall problem into smaller subproblems. Instead, the complete airline scheduling problem is solved at once. Because even smaller subproblems are *NP*-hard, the complete problem is computationally intractable with standard exact solution algorithms. Thus, to solve this problem without an improper or unacceptable simplification and reduction of the problem, metaheuristics have to be used.

Metaheuristics iteratively improve solutions to a given problem (see Sect. 3.2). For the airline scheduling problem, a complete airline schedule is its solution, which then has to be processed by a metaheuristic. This schedule implicitly includes all partial solutions of the traditional approach including their interdependencies. Processing complete airline schedules results in a truly simultaneous planning approach.

In the next section, the conceptual design of the simultaneous approach is presented. It focuses on the four basic design elements of metaheuristics that have to be designed to match the given problem. Also in this section, three metaheuristic techniques are presented as representative examples for different search strategies: threshold accepting for local search, a selecto-recombinative genetic algorithm for pure recombination-based search, and a genetic algorithm with both, local and recombination-based search. Each metaheuristic is applied to the airline scheduling problem in Sect. 4.4.3. Sect. 4.4.3.1 presents calibration results for the three different solution techniques. In addition, based on results from the calibrated models, one metaheuristic is selected for further analysis in Sect. 4.4.3.2 and is compared to the sequential planning approach.

4.4.2 Conceptual Design

4.4.2.1 Overview

In Sect. 3.2, four basic elements were identified that need to be addressed when applying a metaheuristic to a specific problem:

1. representation and variation operators,
2. fitness function,
3. initialization,
4. search strategy.

In the following, the design of these elements to simultaneously solve the airline scheduling problem is presented.

4.4.2.2 Representation and Variation Operators

A representation determines the mapping between a phenotype and a genotype. The phenotype is the real solution of the given problem, the genotype represents this solution within a metaheuristic and is subject to the heuristic's variation operators. Thus, the representation and the operators work together and cannot be developed independently. An indirect representation for the airline scheduling problem is proposed in the next section, followed by the introduction of proper operators. Finally, although not elementary for metaheuristics in general, repair operators are presented that are necessary to deal with infeasible solutions.

Representation. A complete airline schedule is a solution of the airline scheduling problem. Thus, each genotype or individual[22] has to encode decision variables that – combined with given data – form a complete airline schedule. This schedule has to include the set of flights that are carried out including the assignment of the available resources.

One of the most important restrictions when scheduling a flight is the availability of an aircraft of the desired fleet type at the planned departure time and departure airport. The representation concept presented in the following has been designed to incorporate these restrictions, it only allows the representation of flights that can be carried out with the available aircraft. Thus, each encoded solution is feasible with respect to the aircraft assignment and ensures the satisfaction of flight coverage, flow balance, and aircraft count constraints.

Each genotype consists of a fixed number S of segments. This number corresponds to the number of aircraft available, and each segment represents the LOF of one day for one aircraft. The segments are ordered in succession according to the fleet type, i.e. all segments of one fleet type are connected to each other. Each segment contains a sequence of tuples. Let L_s denote the number of tuples in the segment s and l_s the position in the segment. Each tuple consists of an airport a_{l_s} and a time t_{l_s}, indicating the location and duration the aircraft is resting on ground. Fig. 4.36 presents an overview of this representation concept with seven aircraft of two different fleet types. The aircraft is required to fly between the encoded airports in the indicated sequence, thus, there is an indirect encoding of the flights depending on the ground activities. Using this indirect and aircraft-based representation avoids infeasible aircraft assignments.[23]

The departure time of a flight depends on the departure times of the preceding flights. The airport a_{l_s} at position l_s is the departure airport of the flight f_{l_s} connecting airports a_{l_s} and a_{l_s+1}. The encoded time interval t_{l_s} at each position in the segment indicates the amount of time that the aircraft is scheduled to remain on the ground after the minimum turn time has elapsed and before conducting the next flight. The departure time $t_{l_s}^{dep}$ of flight f_{l_s} can be calculated recursively based on

---

[22] The terms genotype and individual are used synonymously in the following.

[23] The daily model can easily be extended in future work to a weekly problem by constructing seven LOFs per aircraft. However, complexity will also significantly increase.

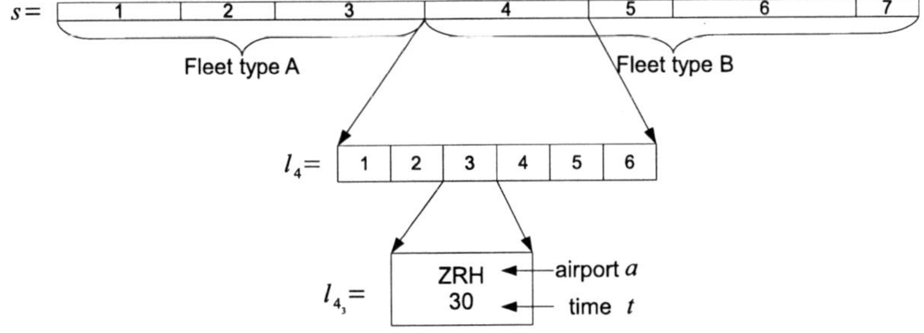

Fig. 4.36 Representation concept

DUS	OSL	DUS	ZRH	DUS	VIE	DUS	ZRH
10	10	30	20	20	0	10	10

Fig. 4.37 Segment with flight information (genotype) for one aircraft

the departure time of the previous flight $t_{l_s-1}^{dep}$, its block-time $t_{a_{l_s}-1,a_{l_s}}^{block}$, the minimum required turn-time $t_{a_{l_s}-1}^{turn}$ and the scheduled ground time t_{l_s}:

$$t_{l_s}^{dep} = t_{l_s-1}^{dep} + t_{a_{l_s}-1,a_{l_s}}^{block} + t_{a_{l_s}-1}^{turn} + t_{l_s}. \tag{4.55}$$

If a departure time is scheduled when night flying restrictions at the corresponding airport are in effect, the first point in time flights are allowed is used as a point of reference instead of the ready time of the previous flight. This is also the procedure to calculate the first departure time t_1^{dep} of the segment, because this cannot be calculated recursively. If on the other hand the airport allows flight operations throughout the night, any point in time can be chosen as point of reference for the departure time calculation of t_1^{dep}.

Using the decision parameters (a and t) and the given data (t^{block}, t^{turn}, and night flying restrictions), a flight schedule can be determined.

As an example, the segment from Fig. 4.37 (one aircraft) is decoded using the following data:

- night flying restrictions at DUS from 23:00 to 06:00 (UTC),
- minimum turn times $t^{turn} = 30$ minutes at all airports,
- block times of flights:

$$t_{DUS,OSL}^{block} = t_{OSL,DUS}^{block} = 110,$$
$$t_{DUS,VIE}^{block} = t_{VIE,DUS}^{block} = 80, \text{ and}$$
$$t_{DUS,ZRH}^{block} = t_{ZRH,DUS}^{block} = 80.$$

In this example, the aircraft departs DUS at 06:10 (airport opens at 06:00, $t_1^{ground} = 10$), arrives in OSL at 08:00 ($t_{DUS,OSL}^{block} = 110$), and departs for the second flight of this segment heading to DUS at 08:40 ($t_{OSL}^{turn} = 30, t_2^{ground} = 10$) etc. Based on the given individual from Fig. 4.37, the schedule in Fig. 4.38 can be determined.

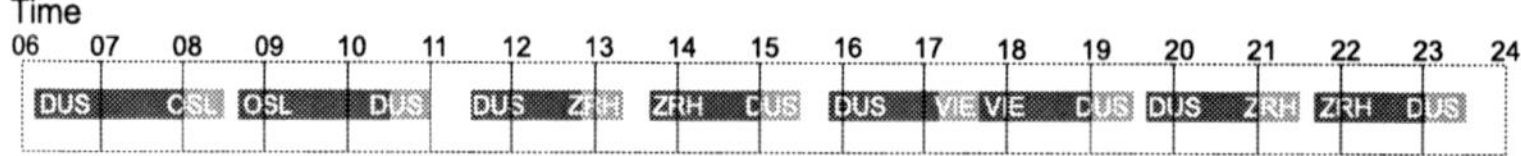

Fig. 4.38 Schedule for one aircraft (phenotype) represented by segment in Fig. 4.37

In this example, a single segment is considered, and the aircraft's last flight is heading back to the airport where the first flight of the segment departed. If more than one aircraft of the same fleet type exist, their segments are joined together and the last flight of each segment is heading towards the starting airport of the next segment. This procedure leads to one single rotation per fleet type, and each aircraft of the same fleet type accomplishes the identical routing consisting of a number of days equal to the number of aircraft in the fleet. Although many airlines favor having only one rotation per fleet because of evenly distributed use of the aircraft and ease of scheduling (Gu et al., 1994, Clarke et al., 1997), some airlines consider more than one rotation for each fleet. To represent these cases, a flag r_s is introduced to each segment s indicating the termination ($r_s = 1$) or continuation ($r_s = 0$) of the rotation. If $r_s = 0$, the last flight of segment s is heading towards the first airport of the succeeding segment if this segment belongs to the same aircraft type. If $r_s = 1$, the last flight of s is heading towards the first airport of the segment that the actual rotation was started with. Fig. 4.39 with four aircraft illustrates the concept of the rotation flags. The individual represents two rotations with two aircraft and a length of two days each. In the first rotation, the last flight of the first day is from DUS to MAD. On the second day, the last flight is heading from ZRH towards NCE (instead of FRA), because the rotation flag is set to $r_2 = 1$. Fig. 4.40 presents the resulting schedule.

Variation Operators. The variation or search operators are used by a metaheuristic to produce new solutions. They work with the genotypes and can be divided into two groups: operators that create new solutions by modifying a current solution (local search operators) and operators that construct new solutions by recombining parts of two or more solutions (recombination-based operators).

Local Search Operators. Local search operators construct a new solution by applying small changes to the current solution. Thus, a local search operator should produce a solution that is in the original solution's neighborhood in the search space, keeping most of the original solution's properties. Based on the given problem here and the representation used, there are solution elements of a different scale in each

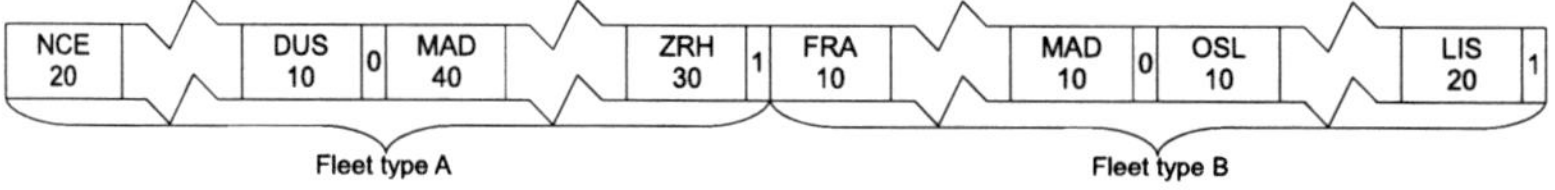

Fig. 4.39 Segment with flight information for four aircraft including rotation flags (genotype)

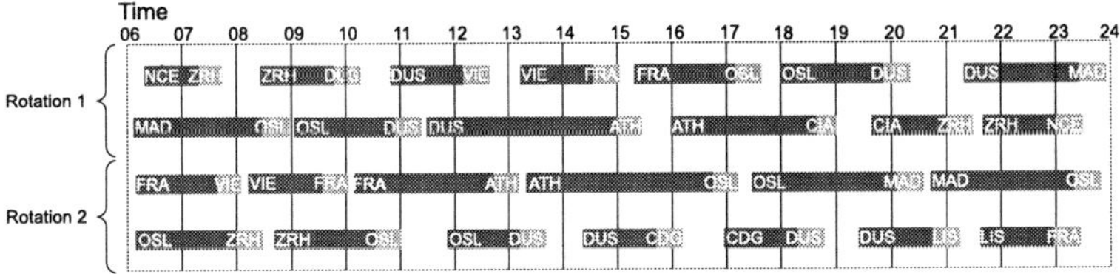

Fig. 4.40 Schedule for four aircraft (phenotype) represented by individual in Fig. 4.39

Fig. 4.41 Deletion of ground time (local search operator *locDelGT*)

genotype (airports, times, aircraft assignments, etc.). A neighboring solution is obtained by modifying one element of the genotype per search step. The search step itself has to address the different elements of a solution accordingly, and even when modifying the same type of element, there are various options to perform the modification. In this study, different operators are developed that can all be applied to the genotypes and can be used complementary and exchangeable. To decide on a final set of operators based on their quality and their impact on the search process's efficiency, additional experiments need to be conducted. Another approach is to develop an adaptive implementation of the operators that automatically applies those operators leading to the best solution quality. This approach is followed in this study. The adaptive implementation is described when presenting the search strategy on page 140.

In the following, the different local search operators developed for the simultaneous airline scheduling approach are presented, each introducing a different type of neighborhood.

- Delete Ground Time (*locDelGT*)
  A genotype encodes a sequence of airports and scheduled ground times at each airport. This local search operator chooses a random segment s of the genotype and a random position l_s within this segment and sets the corresponding ground time $t_{l_s} = 0$. The resulting airline schedule then contains flights f_{m_s} with $m_s >$ l_s with earlier departure and arrival times than the original solution. Fig. 4.41

illustrates the operator on the genotype and the corresponding changes in the phenotype.

- Insert Ground Time (*locInsGT*)
  This operator works vice versa to *locDelGT*: The ground time t_{l_s} of a random position l_s in a randomly chosen segment s is increased by a time parameter t^{init}. All flights following the encoded position are displaced by t^{init}.
- Change Airport (*locChgApt*)
  This operator randomly changes the airport a_{l_s} of a random position l_s in a randomly chosen segment s. The operator results in two different flights, since a_{l_s} represents an arrival and a departure airport of two succeeding flights. In addition, if the block time of the new flights is different to the preceding block time, the flights following these new flights are shifted forward or backwards. Fig. 4.42 illustrates the operator on the genotype and the corresponding changes in the phenotype.
- Delete Airport (*locDelApt*)
  The airport a_{l_s} and the ground time t_{l_s} of a random position l_s in a randomly chosen segment s are removed from the genotype. This operator replaces the two flights connected via a_{l_s} with one flight from a_{l_s-1} to a_{l_s+1}. In addition, because the single new flight probably has a lower block time than the cumulated block time of the two original flights, all flights following l_s in the segment will depart earlier. See Fig. 4.43 for an example of this operator.
- Insert Airport (*locInsApt*)
  This operator works vice versa to *locDelApt*. It inserts a new airport a_{l_s} and ground time t_{l_s} at a random position l_s in a randomly chosen segment s, replacing one flight with two new flights connecting via a_{l_s}. The new ground time is chosen randomly between 0 and the parameter t^{init}. The flights following the position l_s will probably depart later than in the original solution because of the extended cumulated block time of the new flights.
- Change Rotation Flag (*locChgRot*)
  This operator switches the flag r_s of a randomly chosen segment of the genotype. This changes the routing of the affected aircraft and thus their utilization.

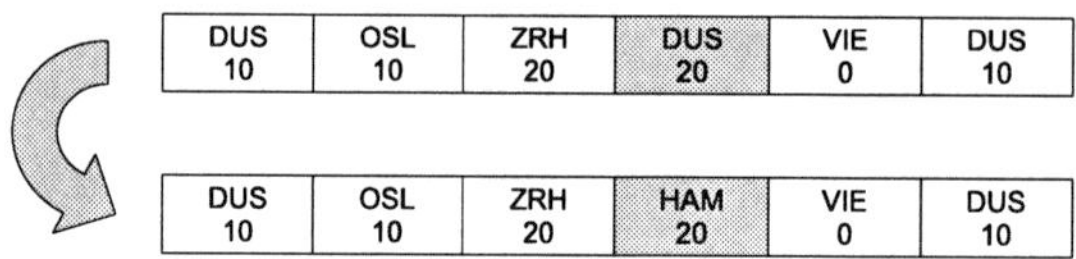

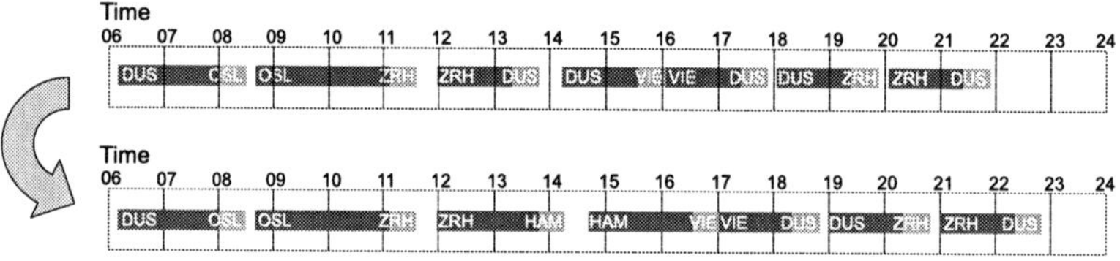

Fig. 4.42 Change of airport (local search operator *locChgApt*)

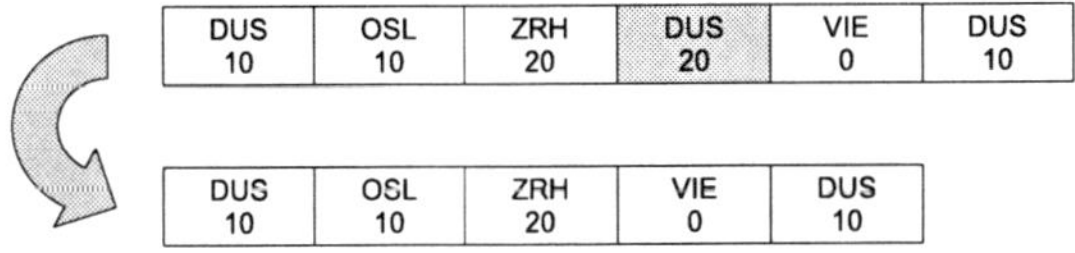

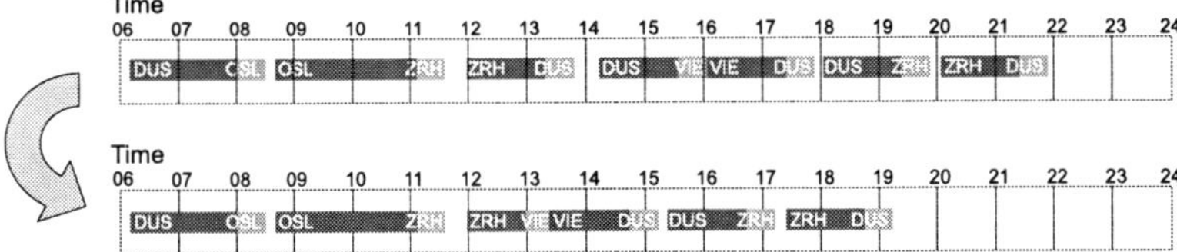

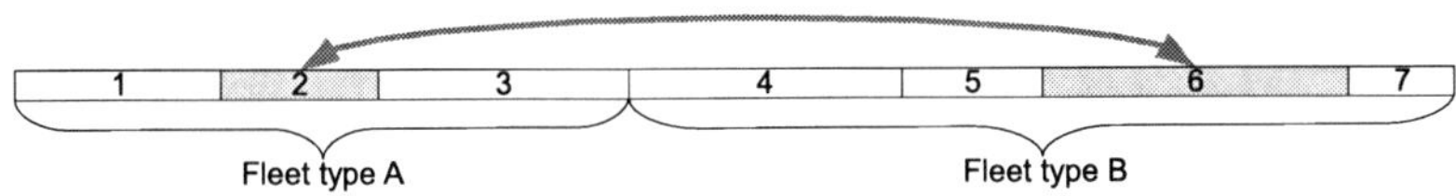

Fig. 4.43 Deletion of airport (local search operator *locDelApt*)

Fig. 4.44 Change of fleet assignment (local search operator *locChgFA*)

In addition, because either two rotations are joined together or a single rotation
is split up, the flights at the connecting points are rerouted.

- Change Fleet Assignment (*locChgFA*)
  This operator changes the order of the segments in the genotype. Because the
  segments of one fleet type are grouped together, exchanging two segments s_1
  and s_2 changes the fleet assignment if the two segments belong to different fleet
  types. In addition to a different fleet assignment, the flights at the connecting
  points between the segments might be changed if the segment is part of a multi-
  segment rotation. Fig. 4.44 illustrates this operator.

- Change Airports with Similar Market Size (*locChgAptMS*)
  The operator *locChgApt* randomly changes the airport at a random position l_s
  of a randomly chosen segment s. The choice among the different airports is
  discrete, thus, a neighboring solution could contain any other airport at the
  specified position. To increase the locality of the search space with this type
  of operator, similarities between different airports have to be defined. The operator *locChgAptMS* presented here uses the total market size of an airport as
  a characteristic to measure similarity. The smaller the difference between two
  market sizes of two airports, the higher their similarity. Given the market size
  ms_{od} for every market of an originating airport $o \in A$ and a destination airport
  $d \in A$, the total market size ms_a of airport $a \in A$ is calculated as:[24]

$$ms_a = \sum_{d \in A} ms_{ad} + \sum_{o \in A} ms_{oa}. \tag{4.56}$$

---

[24] Market sizes are given by the market size estimation model presented in Sect. 4.2.2.

The probability p_{ab} of an airport b to replace an airport a in the genotype by operator *locChgAptMS* increases with decreasing difference in their market sizes and is calculated as follows:[25]

$$p_{ab} = \frac{(max(|ms_a - ms_b|, 1))^{-1}}{\sum\limits_{c \in A} (max(|ms_a - ms_c|, 1))^{-1}}. \tag{4.57}$$

- Change Airports with Similar Distance (*locChgAptDist*)
  This operator works in analogy to *locChgAptMS*, except it uses the geographical distance $dist_{ab}$ as indication of similarity of airports a and b. The closer two airports, the higher the probability that they are exchanged in the genotype. The probability p_{ab} of an airport b to replace an airport a is calculated as:

$$p_{ab} = \frac{(max(dist_{ab}, 1))^{-1}}{\sum\limits_{c \in A} (max(dist_{ac}, 1))^{-1}}. \tag{4.58}$$

Recombination-based Operators. The purpose of recombination-based operators is to create new solutions by combining meaningful elements of different solutions. A population of solutions is necessary because each new solution (*offspring*) emerges from at least two preceding solutions (*parents*). The recombination-based or crossover operators work well on decomposable problems which contain subproblems that are quasi-independent and could be solved separately. Then, the solution process consists of combining good partial solutions to construct an overall good solution.

Different standard recombination operators were developed that work on string genotypes. Examples are 1-point, n-point, or uniform crossover. However, although string-type genotypes are used here, in contrast to the traditional string representation there is no fixed length of the genotype. The number of stops in a segment depends on the block times of the flights and the scheduled ground times, and this number changes during optimization. Thus, the length of a genotype (in terms of encoded stops) is variable, even if the number of segments remains constant. Since the standard operators work on fixed-length strings, they cannot be directly used here but may be adapted to match the variable-length genotypes.

In the following, different recombination operators are presented. As for the local search operators, there is an adaptive control of their application during the search process, which is presented on page 140.

- 1-Point Recombination (*rec1P*)
  The traditional 1-point crossover splits two genotype strings at a random position and exchanges the partial strings. For the representation used in this approach, using the absolute position l in a genotype could result in substrings with a large time displacement. If for example one parental solution contains many short flights and the second parent only long flights, the same absolute

---

[25] The *max*-function is required to avoid the division by 0.

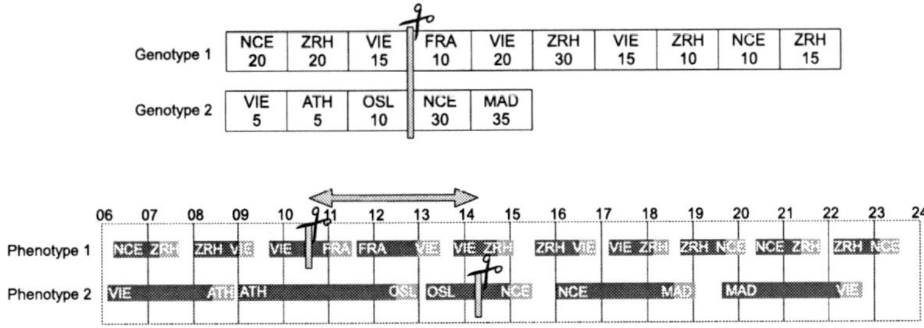

Fig. 4.45 Recombination using an absolute crossover position

position would indicate a flight of the first parent departing much earlier than the corresponding flight of the second parent. This example is illustrated in Fig. 4.45. In this approach, instead of using the absolute position, a segment s and a time t^{cross} are used to indicate the crossover point. s and t^{cross} are chosen randomly. Then, the absolute position within the genotype is determined based on s and t^{cross} independently for each parent. This will ensure the extraction of sub-strings of approximately the same time interval of two individuals and reduce displacement of flights after the crossover position. The recombination itself is conducted like a standard 1-point crossover. At the crossover point, a flight is exchanged, since the arrival airport is changed. In addition, since a new sequence of flights is introduced to a rotation, a second flight is changed to ensure connectivity of the rotation.

In the following figure, a simple recombination operation is illustrated with an example ($t^{cross} = 14{:}30$).

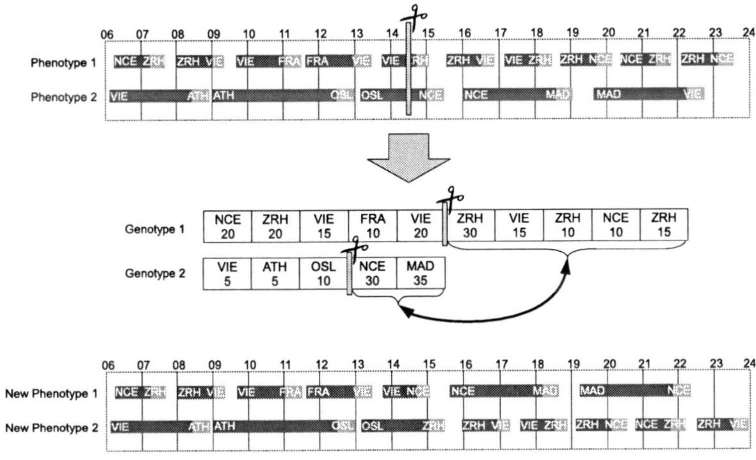

Fig. 4.46 Recombination using a time-dependent crossover position (recombination operator *rec1P*)

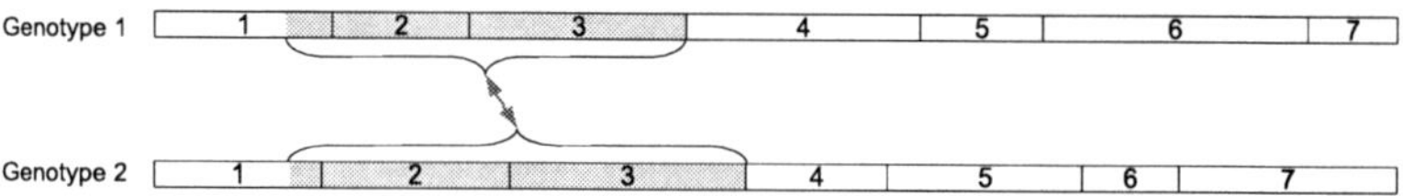

Fig. 4.47 Recombination using two time-dependent crossover positions (recombination operator *rec2P*)

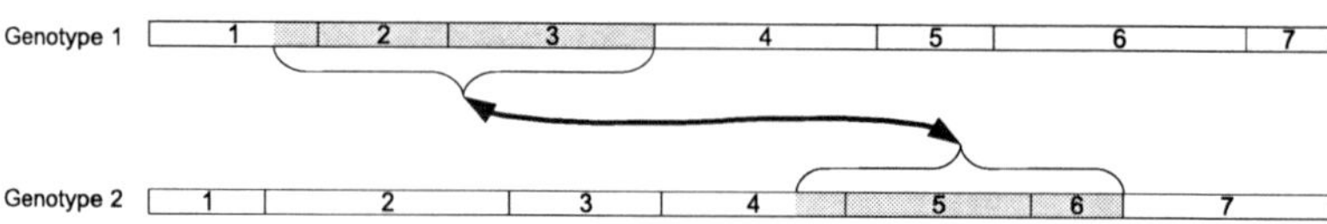

Fig. 4.48 Recombination of substrings (recombinationoperator *recString*)

- 2-Point Recombination (*rec2P*):
  The operator *rec2P* works similarly to *rec1P* besides randomly choosing an additional time t^{cross2} defining a second crossover position ($t^{cross} + t^{cross2}$). Thus, each genotype is divided into three substrings, and the inner string is exchanged between the parents. An illustrative example is given in Fig. 4.47.
- String Recombination (*recString*)
  With *rec2P*, elements of the genotype are exchanged between approximately the same positions. The operator *recString* works similarly to *rec2P*, however, the segment s_1 at one parent where the substring of the other parent is inserted does not need to correspond to this segment s_2 ($s_1 \neq s_2$). This operator allows the change of the fleet assignment. The example in Fig. 4.48 illustrates the operator *recString*.

Repair Operators. The representation chosen allows the encoding of only feasible aircraft assignments: because of the aircraft-based relative encoding, each flight is covered by exactly one aircraft, the number of aircraft per fleet is not exceeded, and the flow balance is preserved. In addition, the search operators consider operational restrictions of the fleets, for example they do not insert an airport in a segment that the corresponding fleet type is not able to operate or that results in a flight being too long for the aircraft's endurance. However, other restrictions exist that might still be violated when applying the search operators.

If there are too many stops in one segment, the last flights might be scheduled to depart on the next day, which is encoded in the subsequent segment. If the search operators include new stops or replace flights by flights with longer block times, the subsequent flights are postponed because of the relative encoding of their departure times. If the end of day is exceeded by a segment, its elements are checked for scheduled ground time that can be minimized. If there is not sufficient ground time available, randomly chosen stops are deleted until the resulting flight program can be completed in one day.

Any night flight restrictions are always met for the departure of a flight, since the departure times are encoded in relation to either the ready time of the previous flight or the time an airport allows flight operations. However, the arrival of a flight might violate some restrictions, if for example a flight departed from an airport without night restrictions heading to an airport with restrictions. If a flight arrives too early, it is delayed by extending the scheduled ground times. If the flight arrives too late in the evening, the scheduled ground times are minimized or - if minimizing is not sufficient - stops are randomly deleted from the segment.

4.4.2.3 Fitness Function

An individual's fitness corresponds to the quality of the encoded solution. The quality is quantified by the objective value of the solution. The objective of the airline scheduling problem is to construct airline schedules with maximum operating profit, thus, this measure is used for fitness evaluation. The operating profit is calculated using the schedule evaluation procedure presented in Sect. 4.2. In contrast to the objective function, the fitness function of a metaheuristic can contain additional properties of a solution that are not expressed by the objective value. One example is constraints of the problem and possible violations of the solution. Then, the fitness value of the solution considers additional penalties that describe the existence of constraint violations. If possible and practicable, the penalty costs should reflect the extent of the violation to establish locality in the search space with respect to the feasibility of the solution. Setting the amount of penalties represents a difficult task, since there is a trade-off between a too restrictive search because of too high penalty costs and the possibility of obtaining an infeasible solution at the end of the optimization because of too low penalty costs.

The selected representation and operators (including the repair operators) result in feasible solutions of the airline scheduling problem except for maintenance considerations. Each aircraft is required to undergo maintenance checks at regular intervals (for example every three days) at maintenance stations. In this study, the maintenance restrictions are included as penalty function: if an aircraft is not scheduled for appropriate maintenance, penalty costs reduce the fitness value by a certain amount controlled via a parameter. These costs increase with the number of aircraft violating the maintenance restrictions and the delay of the required maintenance for each aircraft. Quantifying the penalty costs is easy since a solution contains an airline schedule which can be easily checked for maintenance opportunities for every aircraft and the time in-between. A maintenance opportunity is given if an aircraft is on the ground at a maintenance airport for the required time necessary to perform the maintenance checks and the maintenance station has enough capacities left for this specific aircraft type.

4.4.2.4 Initialization

Because metaheuristics represent problem-specific improvement heuristics, a solution (or a population of solutions) has to be created at the start of the search process.

If there is problem-specific information and knowledge of high-quality solutions, the search could already focus on good regions of the search space by initializing high-quality solutions. In this study it is assumed that there is not such knowledge. In this case, every solution of the search space should have the same probability of being selected as an initial solution. This is accomplished by randomly choosing among the decision variables of the solution. Thus, when initializing a genotype, a randomly chosen airport is included into each segment until the corresponding LOF reaches the end of day. The ground times t assigned to each stop are chosen uniformly between 0 and a parameter t^{init}.

4.4.2.5 Search Strategy

Two basic search strategies can be distinguished following the main search operators: local and recombination-based search. The choice between both strategies has to be made specifically for each problem. Local search should be used if the locality of the problem is high, because then the structure of the search space can guide the search towards high-quality regions. In contrast, if the problem is decomposable, recombination-based operators can be applied that solve subproblems of the given problem independently and combine the good partial solutions to good overall solutions. However, identifying locality or decomposability of a real-world problem is often difficult, because most problems have both properties. Thus, either experiments on the problem have to be conducted to identify the more relevant concept, or a technique incorporating local search and recombination-based search has to be used. In this study, both approaches are presented and three metaheuristic techniques are implemented:

1. threshold accepting (TA) as a representative example of local search,
2. a selecto-recombinative steady-state genetic algorithm (rGA) as an example of recombination-based search, and
3. a standard steady-state genetic algorithm (GA) as an example of metaheuristic search with both, local and recombination-based search.

Even when focusing on a single search strategy (local or recombination-based), an explicit control of the search operators is necessary, because for each type of search operator different variants were designed. In each search step, one operator has to be selected. This selection can follow a given rule, for example each operator is used one after another. Another approach would be to select the operator in each search step randomly. In this study, an adaptive control of the operators is developed which applies those operators in each search step that were advantageous in the previous steps. This procedure not only reduces the number of decisions and parameters required to be set manually, but also increases the efficiency of the search process. The adaptive control used in this approach randomly selects one operator per search step, the probability of each operator to be selected depends on its contribution to the past search progress. If the application of the operator resulted in high-quality solutions (compared to the results of other operators), its selection probability is increased. For each operator $o \in O$ of the current search type, the progress of its last N applications is monitored. Its progress is evaluated according to the change in the fitness value $f_n^o = f(s_o^*) - f(s)$ between the original solution s and solution s_o^* resulting from the nth application of the operator o. The relative fitness contribution c_o of operator o is calculated as:

$$c_o = \frac{max\left(\sum_{n \in N} f_n^o, 0\right)}{\sum_{q \in O} max\left(\sum_{n \in N} f_n^q, 0\right)}. \tag{4.59}$$

Based on this fitness contribution, the selection probability p_o of operator o is calculated as:

$$p_o = \frac{c_o}{\sum_{q \in O} c_q}. \tag{4.60}$$

To prevent diminishing operators, each operator has a minimum selection probability of 0.05. The initial setting for the selection probability of each operator is determined by applying the operator N times to an initial solution without replacement by the obtained new solution.

This adaptive procedure is also applied to the standard steady-state GA to choose among local and recombination-based search. Thus, the GA uses a two-step adaption: first, the type of operator (local or recombination-based) is chosen, then, the operator itself is selected. The probability of selecting a recombination-based operator depends on the average progress contribution of all recombination-based search operators. Let R denote the set of recombination-based operators and L the set of local search operators ($O = R \cup L$). The probability p_R of using a recombination-based search in the current search step is then calculated as:

$$p_R = \frac{\dfrac{\sum_{r \in R} c_r}{|R|}}{\dfrac{\sum_{r \in R} c_r}{|R|} + \dfrac{\sum_{l \in L} c_l}{|L|}}. \tag{4.61}$$

Algorithms 7, 8, and 9 present the final specifications of the TA, rGA, and GA including their parameters.

4.4.3 Experiments

In this section, results from implementing and applying the simultaneous airline scheduling approach are presented. First, calibration results focusing on the setting of the parameters are presented. In addition, a second objective is to decide among the three different solution strategies. Using the strategy resulting in the best solutions, the search process and the obtained solutions are analyzed.

Algorithm 7. Threshold Accepting

1: choose parameters:

2:          initial threshold $T \in [0,1]$

3:          threshold reduction step size $r < T$

4:          maximum number of iterations $i_{decrease}$ between threshold reduction

5:          maximum number of iterations $i_{max} > i_{decrease}$ when $T = 0$

6: create initial solution s with fitness value $f(s)$

7: calculate p_o for all operators $o \in O$

8: iteration $i = 0$

9: **repeat**

10:    $i = i + 1$

11:    select local search operator $o \in O$ according to p_o

12:    create neighboring solution s_o^*

13:    calculate new fitness value $f(s_o^*)$

14:    $\Delta f = f(s) - f(s_o^*)$

15:    **if** $\Delta f < (T \cdot f(s))$ **then**

16:       $s = s_o^*$

17:       update p_o for all operators $o \in O$

18:    **end if**

19:    **if** $T > 0$ and $i > i_{decrease}$ **then**

20:       $T = T - r$

21:       $i = 0$

22:    **end if**

23: **until** $i = i_{max}$

The metaheuristics have been implemented in C++. The experiments presented here were conducted on different workstations with different processor and memory specifications.

Because a major goal of this study is to compare the sequential and simultaneous airline scheduling approach, both approaches should have equal requirements and capabilities. To allow a fair comparison, some slight modifications to the conceptual design of the metaheuristic are applied. One modification aims at the rotation encoding in each genotype. Because the sequential approach only allows one rotation per fleet type, the rotation flags are set to $r_s = 0$ for all S and the local search operator *locChgRot* is not allowed to change the flags. Furthermore, the sequential approach does not consider any capacity constraints at maintenance stations; in theory, all aircraft can be scheduled to undergo maintenance on the same night. In addition, there is no consideration of a minimum time necessary to perform the maintenance. This less restrictive specification is applied to the calculation of maintenance penalty costs in the simultaneous approach. On the other hand, because the

---

Algorithm 8. Selecto-Recombinative Steady-State Genetic Algorithm

---

1: choose parameters:

2:         population size n

3:         p_{conv} to determine convergence of the population

4: create initial population S_0 with n numbers of solutions s

5: calculate fitness value $f(s)$ for each $s \in S$

6: calculate p_o for all operators $o \in O$

7: iteration $i = 0$

8: **repeat**

9:    $i = i + 1$

10:    select recombination-based search operator $o \in O$ according to p_o

11:    choose two solutions s_1 and s_2 randomly

12:    create solution s_o^* from s_1 and s_2

13:    calculate new fitness value $f(s_o^*)$

14:    replace the worst solution in S by s_o^*

15:    update p_o for all operators $o \in O$

16:    determine solution $\hat{s} \in S$ with maximum fitness

17:    calculate average fitness $\bar{f}(S)$ of population

18: **until** $f(\hat{s}) - \bar{f}(S) < p_{conv} \cdot f(\hat{s})$

---

sequential approach always fulfills the maintenance constraints, the metaheuristics penalty costs are set to a high value (500,000) to de facto ensure meeting the three-day maintenance requirements.[26]

4.4.3.1 Calibration

One advantage of the adaptive control of the search operators is the reduction of the number of parameters to be set for each metaheuristic. However, each metaheuristic presented previously is still controlled by some parameters that have to be calibrated. In this section, experiments on different parameter settings are presented to find a final calibration for each metaheuristic. Then, the solution quality of all three calibrated models is compared to decide about the search strategy (TA, rGA, or GA) for further analysis and comparison with the sequential airline scheduling approach.

The calibration process is conducted in analogy to the calibration of the sequential approach (see Sect. 4.3.4.1). Different parameter settings are examined by applying each metaheuristic to five different planning scenarios (see Appendix B). For each parameter setting and scenario, five runs are conducted. The results presented in the following thus represent averages of these five runs. In analogy to the results

---

[26] Additional parameters are $N = 5$ for the adaptive control of the search parameters and $t^{init} = 25$ for initialization and some search operators.

Algorithm 9. Standard Steady-State Genetic Algorithm

1: choose parameters:

2:          population size n

3:          p_{conv} to determine convergence of the population

4: create initial population S_0 with n numbers of solutions s

5: calculate fitness value $f(s)$ for each $s \in S$

6: calculate p_o for all operators $o \in O$

7: calculate p_R and p_L

8: iteration $i = 0$

9: **repeat**

10:     $i = i + 1$

11:     **if** $random(0,1) < p_R$ **then**

12:        select recombination-based search operator $o \in R$ according to p_o

13:        choose two solutions s_1 and s_2 randomly

14:        create solution s_o^* from s_1 and s_2

15:     **else**

16:        select local search operator $o \in L$ according to p_o

17:        create neighboring solution s_o^*

18:     **end if**

19:     calculate new fitness value $f(s_o^*)$

20:     replace the worst solution in S by s_o^*

21:     update p_o for all operators $o \in O$

22:     update p_R and p_L

23:     determine solution $\hat{s} \in S$ with maximum fitness

24:     calculate average fitness $\bar{f}(S)$ of population

25: **until** $f(\hat{s}) - \bar{f}(S) < p_{conv} \cdot f(\hat{s})$

---

from the sequential approach, they include fitness values as a measure of solution quality and the number of fitness evaluations until the best solution was found as a platform-independent quantification of the effort to solve the problem. All values are normalized to aggregate among all scenarios (see page 119).[27]

Threshold Accepting. The TA algorithm was implemented as presented in algorithm 7. This specification uses four parameters to control the search process:

- the initial threshold $T \in [0, 1]$,
- the threshold reduction step size $r < T$,
- the number of iterations $i_{decrease}$ between the threshold reductions, and

---

[27] The individual results of the scenarios including the absolute values are presented in Sect. C.1.2 in the appendix.

- the maximum number of iterations $i_{max} > i_{decrease}$ at the end of the algorithm when $T = 0$ (then the search process represents a local hill climbing algorithm that does not accept inferior solutions).

The impact of the individual parameters is tested by solving the airline scheduling problem with different parameter combinations. In each setting, one parameter is set to different values while the others remain constant. For the constant parameters, the following setting is chosen as the basic setting:

$$T = 0.2,$$
$$r = 0.005,$$
$$i_{decrease} = 20,$$
$$i_{max} = 500.$$

In the following, four diagrams are presented, each illustrating the results for the experiments on one parameter.

Fig. 4.49 presents the calibration of the initial threshold T. The solution quality decreases for low and high values of T, it is highest for values around $0.2 - 0.25$. Thus, these values seem to represent the best compromise between a random search (high T) and a hill-climbing technique that does not accept inferior solutions during search (low T).

Fig. 4.50 presents the calibration of the parameter r. The smaller r, the higher the resulting solution quality. If r is low, the threshold is reduced very slowly, allowing an explorative search. However, the computational effort also increases. Surprisingly, for values of r very close to 0, a decrease in solution quality is observed, which is confirmed by additional experiments on these values.

Fig. 4.51 presents the calibration of the parameter $i_{decrease}$. In general, solution quality increases with higher $i_{decrease}$. The higher $i_{decrease}$, the more search steps are performed before the threshold is further reduced. This allows the exploration of more solution space during optimization. As Fig. 4.51 clearly shows, the higher solution quality is obtained at the cost of increased computational effort.

Fig. 4.49 Aggregated calibration results for parameter T

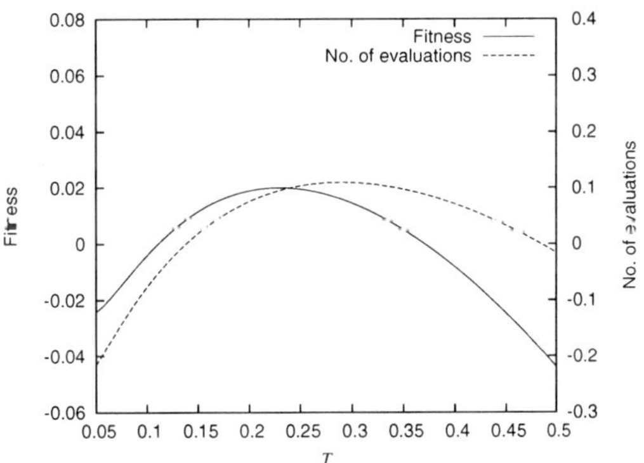

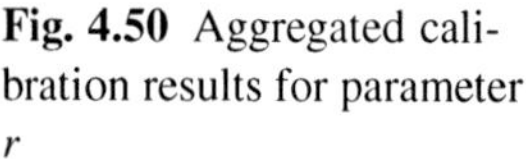

Fig. 4.50 Aggregated calibration results for parameter r

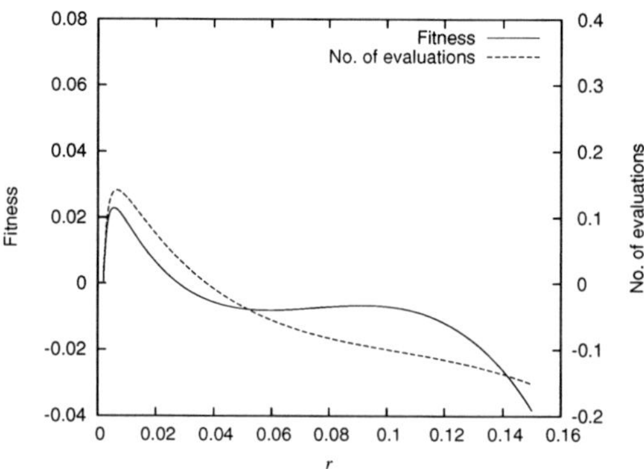

Fig. 4.51 Aggregated calibration results for parameter $i_{decrease}$

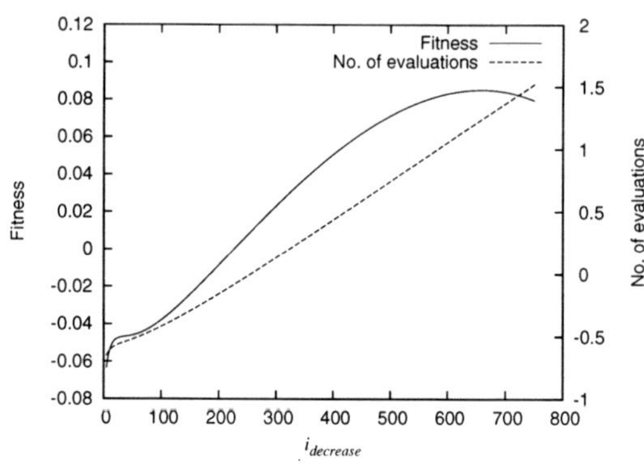

Fig. 4.52 Aggregated calibration results for parameter i_{max}

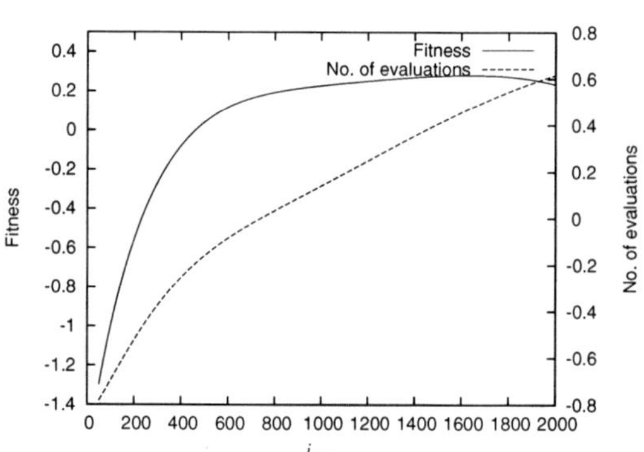

Fig. 4.52 presents the calibration of the parameter i_{max}. This parameter comes into play when the threshold is set to $T = 0$. Then, the TA represents a hill-climbing technique that stops after i_{max} iterations without increase of solution quality. Increasing i_{max} leads to a higher solution quality, since the hill-climbing technique has more attempts to escape from local optima.

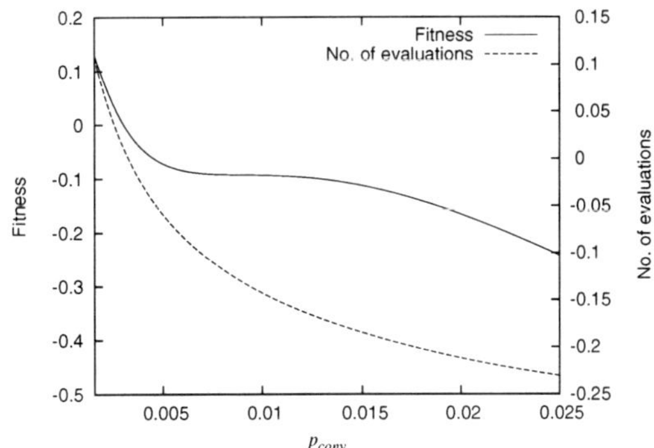

Fig. 4.53 Aggregated calibration results for parameter p_{conv} of the selecto-recombinative genetic algorithm (rGA)

For each parameter, the value at which the best fitness was achieved is used for the subsequent experiments. Thus, the complete final parameter setting is as presented in the following:

$$T = 0.25,$$
$$r = 0.005,$$
$$i_{decrease} = 650,$$
$$i_{max} = 1500.$$

Selecto-Recombinative Steady-State Genetic Algorithm. Algorithm 8 describes the specification of the selecto-recombinative GA. This algorithm uses the following two parameters:

- population size n,
- parameter p_{conv} to determine the convergence of the population to terminate the algorithm. p_{conv} represents a percentage of the fitness value of the best solution, if the difference between the fitness of the best solution and the average fitness of the population is smaller than this value, the algorithm is terminated.

The following two figures 4.53 and 4.54 present the result on different settings for these two parameters. The plotted values are the values of the best solution from every population. The standard setting is chosen as in the following:

$$n = 50,$$
$$p_{conv} = 0.01.$$

Fig. 4.53 presents the calibration of the parameter p_{conv}. The smaller p_{conv}, the higher the required convergence of the population before the algorithm terminates. To reach this convergence, many search steps are necessary, each possibly creating a better solution. This results in overall better solution quality. On the other hand, more search steps require more computational effort. Both effects can clearly be observed in Fig. 4.53.

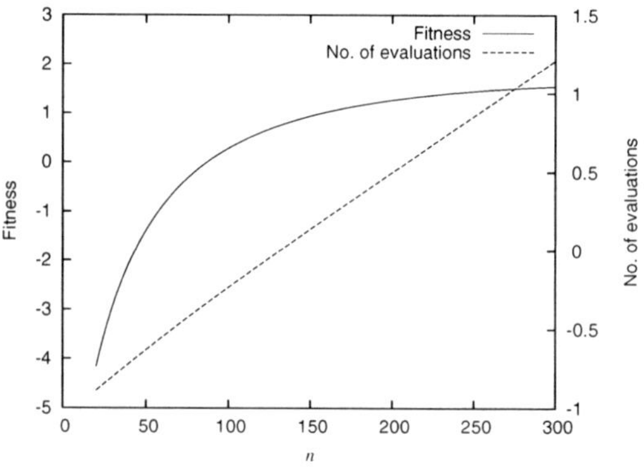

Fig. 4.54 Aggregated calibration results for parameter n of the selecto-recombinative genetic algorithm (rGA)

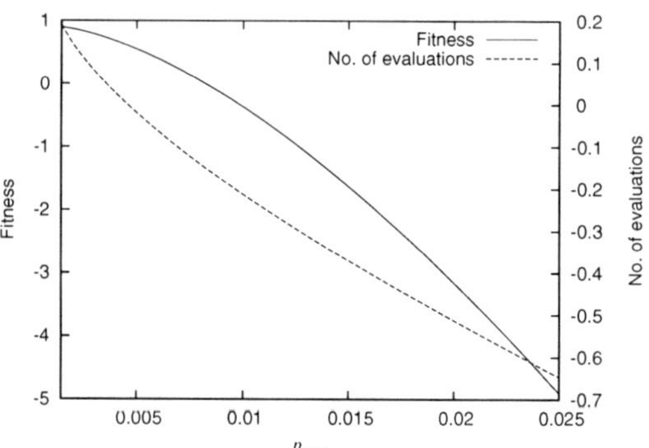

Fig. 4.55 Aggregated calibration results for parameter p_{conv} of the standard genetic algorithm (GA)

Fig. 4.54 presents the calibration of the parameter n. Vice versa to p_{conv}, increasing n yields better solution quality, since more solutions are processed. The more solutions, the higher the chance of the search steps to find a better solution. In addition, convergence is more difficult to achieve because more solutions enter the calculation of the average fitness of the population. As Fig. 4.54 shows, for increasing n the fitness value asymptotically approximates a maximum value, whereas the required number of schedule evaluations constantly increases.

The final parameter setting used in the following for rGA is:

$$n = 200,$$

$$p_{conv} = 0.00125.$$

Standard Steady-State Genetic Algorithm. The specification in algorithm 9 of the standard GA corresponds to the algorithm 8 except for the incorporation of the local search operator. Thus, the parameters are the same as in the selecto-recombinative GA and the same basic setting is used. The following two figures 4.55 and 4.56 present the result for the standard GA on different settings for these two parameters.

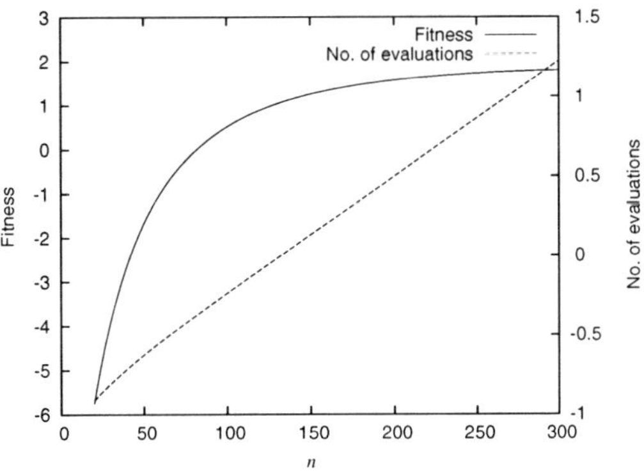

Fig. 4.56 Aggregated calibration results for parameter n of the standard genetic algorithm (GA)

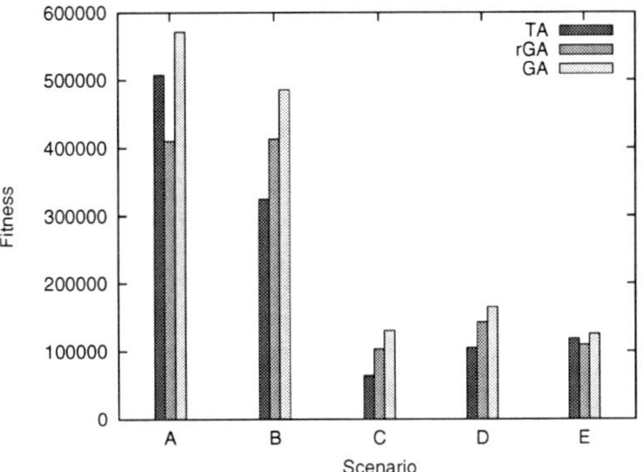

Fig. 4.57 Fitness values for the different search strategies

The results for the standard GA are almost the same as for the selecto-recombinative GA. The final parameter setting for the GA is:

$$n = 200,$$
$$p_{conv} = 0.00125.$$

Strategy Selection. The parameter settings above should lead to high quality solutions for each search strategy. To decide among these strategies, each strategy with its calibrated parameters is applied to the different planning scenarios. The resulting (average) fitness values and number of required fitness evaluations for each strategy and scenario are presented in the following figures 4.57 and 4.58.[28]

For all scenarios, the GA yielded the highest solution quality. Except for scenarios A and E, the selecto-recombinative GA produced better results than TA. Because the GA and rGA use populations of solutions, they require significantly more fitness evaluations than TA, which processes only one solution. These results indicate that a combined local and recombination-based search outperforms search strategies

---

[28] The individual results of the scenarios including the absolute values are presented in Sect. C.1.2.4 in the appendix.

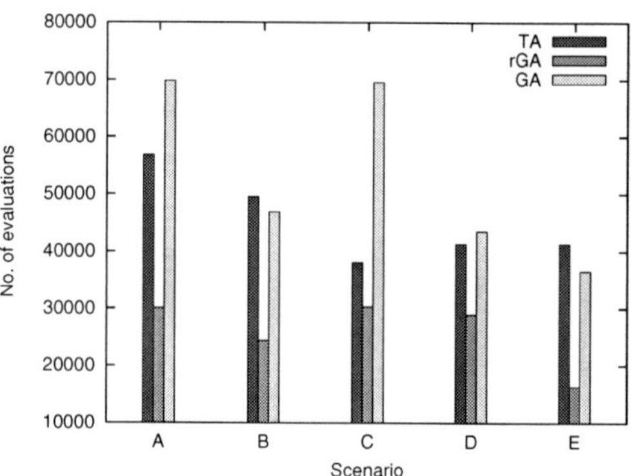

Fig. 4.58 Number of required fitness evaluations for the different search strategies

focusing only on one type of operators. This finding is not surprising, since most problems of practical importance inherit properties applicable to both search concepts, local and recombination-based search (Michalewicz & Fogel, 2000, Rothlauf, 2006a). To validate the results, an unpaired t-test is conducted.[29] The null hypothesis H_0 is that the observed differences in the fitness values are random. H_α says that the differences are a result of the model specification. The critical t-value for $p = 0.975$ is 2.306. The results shown in Table 4.8 for the three models and five scenarios show that the t-values always exceed the critical t-value of the level of significance. Thus, H_0 can be rejected on the 97.5%-level. The GA represents the search strategy that works best using the presented airline scheduling approach.

Table 4.8 t-values for the validation of the search strategy comparison

Models	Scenario				
	A	B	C	D	E
TA vs. rGA	6.405	20.136	13.222	13.637	3.277
TA vs. GA	7.057	79.604	16.775	16.127	2.400
rGA vs. GA	10.554	17.641	5.645	6.251	4.811

4.4.3.2 Analysis

In the following paragraphs, the obtained solutions and the solution process of the GA are analyzed. The following Table 4.9 presents the key figures of the schedules (averaged values, standard deviations in parentheses).[30] Except for the number of evaluations conducted during optimization, the results are very stable. The standard deviations are very low for all key figures of the solutions indicating that the solutions obtained by the different optimization runs for each scenario are similar. Since a random initialization was conducted for each run with an equal selection probability for the decision variables, these results are very satisfactory with regard to

---

[29] The required test of the results for normal distribution was conducted using a Kolmogorov-Smirnov test.

[30] Individual results are presented in tables in Sect. C.2.2.1 in the appendix.

Table 4.9 Key figures of airline schedules constructed with the simultaneous planning approach

Key Figure	Scenario				
	A	B	C	D	E
Profit	571,812	485,775	130,384	165,556	125,880
	(20,556)	(3,086)	(12,019)	(9,894)	(7,466)
SLF	0.344	0.501	0.219	0.495	0.316
	(0.019)	(0.008)	(0.008)	(0.018)	(0.011)
No. of passengers	5,535	4,442	2,551	2,493	1,801
	(113.33)	(88.37)	(68.13)	(64.53)	(224.69)
No. of flights	114	125	70	72	38
	(2.88)	(3.71)	(2.61)	(3.08)	(3.77)
No. of fitness evaluations	69,832	46,909	69,550	43,569	36,580
	(9,168)	(3,196)	(14,477)	(9,626)	(15,899)
Total no. of evaluations	70,723	47,857	71,249	44,721	38,682
	(9,417)	(2,682)	(14,251)	(9,120)	(16,354)

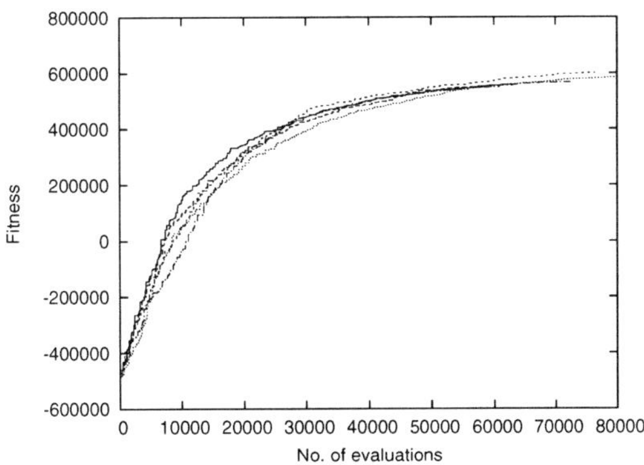

Fig. 4.59 Trend of fitness values of all five optimization runs of scenario A

the stability of the metaheuristic solution approach. Scenario A yielded the highest fitness values, although the seat load factor was best for scenario B. Compared to the uncalibrated genetic algorithm (using the basic parameter setting), the obtained profit values represent an average increase of 52.67%. The number of required fitness evaluations is on average 21.64% higher.

The following figures focus on the solution process. They present results of experiments on scenario A as an representative example for all scenarios.[31] Aggregating the results of the five runs of one scenario or even among the different scenarios would result in meaningless diagrams, since each individual run requires a different number of fitness evaluations leading to different plots of the progress subject to the number of evaluations.

Fig. 4.59 plots the fitness of the best solution in each population. It shows the typical progress for a GA. The progress or improvement of the best solution is highest

---

[31] The results of all scenarios are presented in Sect. C.2.2.2 in the appendix.

Fig. 4.60 Trend of seat
load factors (SLF) of all
five optimization runs of
scenario A

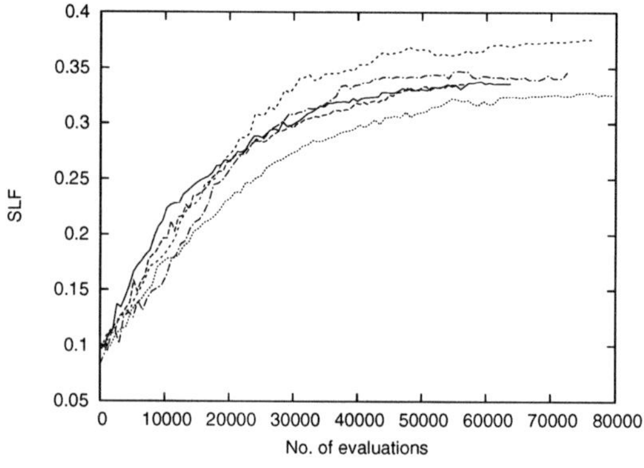

Fig. 4.61 Trend of numbers
of flights of all five optimization runs of scenario A

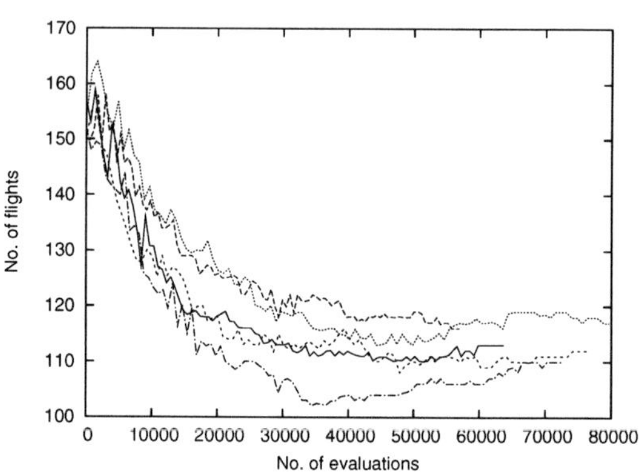

at the beginning of the GA and continuously decreases during optimization. There
are two reasons for this: first, as in every metaheuristic, the search operators have
much room for improvements, since the early solutions inherit random elements due
to the random initialization; second, the population converges during the GA run,
reducing the potential capability of the recombination-based operators with their
rather large modifications to solutions, and leaving only room for local (and, thus,
small) search steps.

Fig. 4.60 presents the SLF for the five runs of scenario A. It depends on the
results in figures 4.61 and 4.62, which plot the number of flights in the best schedule
of each population and the total number of passengers expected to travel on these
flights. Not surprisingly and as indicated by the progress of the fitness values, the
SLF increases during optimization. This increase is a result of an increase in the
number of passengers and a reduction of the number of total flights. Apparently, in
each GA run unprofitable flights are removed from the schedule.[32] Since the number
of flights is higher in the beginning of each run and assuming that the average block
time of all flights remained constant, the final schedules must include some idle

---

[32] Since the fitness function of the GA considers the connectivity of the schedules, the removal
of flights does not depend on their individual passenger demand but also takes their function as
legs of connecting itineraries into account.

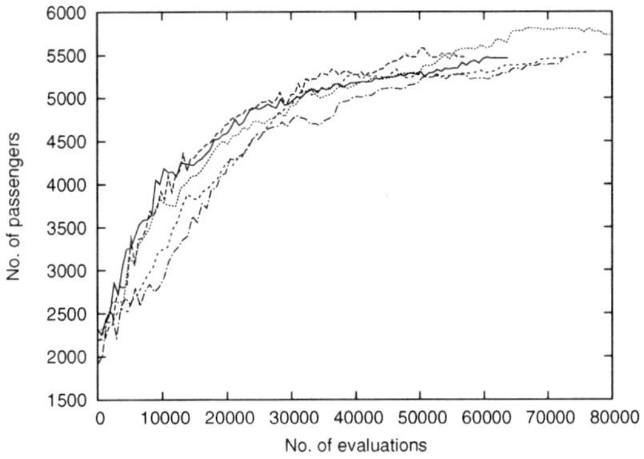

Fig. 4.62 Trend of numbers of passengers of all five optimization runs of scenario A

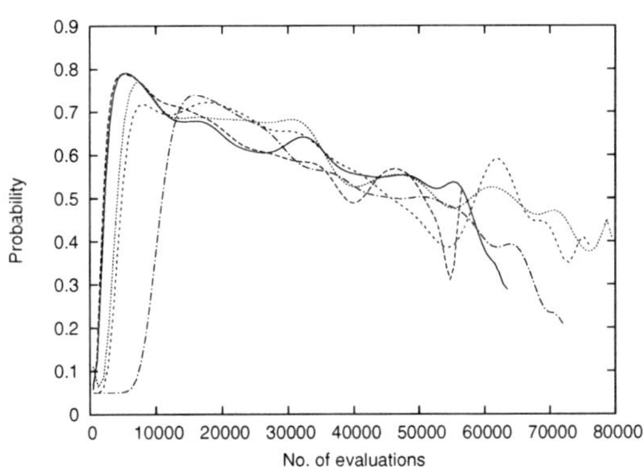

Fig. 4.63 Application probability of recombination-based operators for all five optimization runs of scenario A

ground times. Thus, an additional increase in profit seems possible if the number of airports available for planning is increased to allow the scheduling of additional (profitable) routes. In scenario E this effect can be observed (see Fig. C.26). After removing unprofitable flights, the number of flights in the schedule increases. Since the number of passengers also increases, the SLF remains constant, however, the overall fitness grows.

The GA represents a self-adaptive procedure, thus, another interesting observation is the extent of application of the different search operators during the optimization run. Fig. 4.63 presents the (smoothed) share of the recombination-based operators during the optimization. In general, the different runs show a similar search behavior. The selection probability for recombination-based operators is at its minimum at the start of the optimization. This could be a random result, since in some runs from other scenarios (see Fig. C.28) the initial probability is at the maximum value. However, in all runs recombination becomes the main search operator after the starting phase. Then, its application probability continuously decreases until the end of the optimization. In the final phase of each run, larger variations in the probability exist. The continuous shift from recombination-based to local search can be explained by the different characteristics of the search concepts. Recombination represents a global search operator which is useful for the exploration of

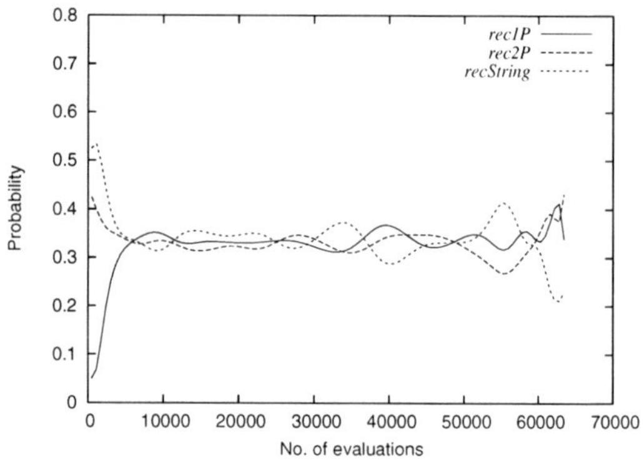

Fig. 4.64 Application probability of the different variants of the recombination-
based search operators
(scenario A, run 3)

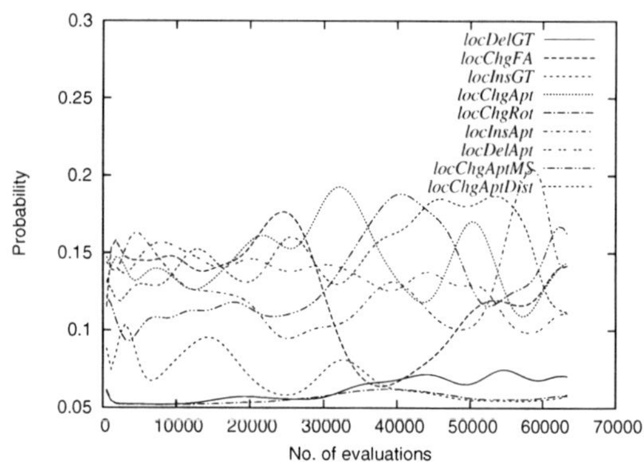

Fig. 4.65 Application probability of the different variants of the local search
operators (scenario A,
run 3)

the search space. The better the solutions of the population and the more the population converges, the more the search has to concentrate on good regions within
the search space (exploitation). This is accomplished by the local search operators.
Then, in the final phase when each operator reaches its limit with regard to solution
improvement, there is no clear advantage of one search concept, leading to the larger
variation of the number of applications of each operator type.

The same variation at the end of the optimization can be observed in Fig. 4.64,
which plots the application of the different variants of the recombination-based
search operators. For clarity, the results of the best run from scenario A are plotted as a representative example.[33] In general, after the initialization the application
probability stabilized at approximately the same value for each operator variant.
Thus, during the optimization each recombination operator is chosen with the same
probability.

Fig. 4.65 presents the application of the different variants of the local search
operator. There is no clear indication of an advantageous variant, since all operators
are used during the optimization to a different and fluctuating extent. However, on

---

[33] Similar figures for the other runs and for the other scenarios are presented in Sect. C.2.2.2 in
the appendix.

average there is a trend that three local search operators are of less priority during search: *locDelGT*, *locInsGT*, and *locInsApt*.

4.4.4 Summary and Conclusion

4.4.4.1 Summary

In this section, a metaheuristic approach for the airline scheduling problem was presented. The four basic elements of a metaheuristic (representation/operators, fitness function, initialization, search strategy) were adapted to solve the airline scheduling problem simultaneously. By representing a complete airline schedule as genotype, the traditional decomposition into smaller subproblems can be avoided. Based on a representation for complete airline schedules, local and recombination-based search operators were developed. As representative examples for different search strategies, three metaheuristic techniques were implemented and calibrated: a TA algorithm as a representative example of local search, a selecto-recombinative GA as an example for recombination-based search, and a traditional GA that incorporates both, local and recombination-based search. The application of the search operators in each technique is controlled adaptively according to the profit contribution of each parameter in past iterations.

Each metaheuristic requires some parameters to control the search process. Values for these parameters are obtained by testing different settings using five different planning scenarios. The individual calibrated set of parameters then consists of those values resulting in the highest solution quality obtained by each heuristic. In a second step, the calibrated models were applied to the five scenarios to decide on the solution strategy. Since the GA produced the solutions with the highest fitness values, this technique was chosen as search strategy for the simultaneous scheduling approach.

Applying the calibrated GA to the five different scenarios yielded very stable results. The standard deviation of some key figures of the obtained schedules among five independent runs for each scenario is very low. Scenario A yielded the best fitness values. The progress of the search process corresponds to theoretical expectations: the increase of fitness is highest at the beginning of the optimization and then asymptotically approaches a maximum value. Because the population converges and its average fitness continuously increases, it is becoming more difficult to find new and better solutions. This trend is confirmed when looking at the extent of the application of the different search concepts of local and recombination-based search. At the beginning of each search, exploration dominates, requiring recombination. With ongoing progress, the exploitation phase gets more and more important, thus, local search operators are applied to a greater extent.

4.4.4.2 Conclusion

One of the major challenges in airline operations research is the integration of the subproblems of the airline scheduling problem into one single model. This has been accomplished with the solution approach presented in this section. By processing complete airline schedules at once, a truly simultaneous airline scheduling is possible. Thus, subproblems of the traditionally decomposed overall problem and their interdependencies are implicitly included. All elements of the schedules are optimized with respect to the overall objective.

In general, in metaheuristic optimization, local and recombination-based search can be distinguished. As most problems of practical relevance include elements that are suitable to both search concepts, those metaheuristics usually work best that use both, local and recombination-based search. This hypothesis is confirmed in this study, since a local search TA and a GA solely based on recombination were outperformed by a GA with both search operators. To further increase the efficiency of the search process, a procedure was developed that adaptively controls the application of the single search operators based on their previous contribution to solution quality. An analysis of their application probability during the optimization run confirms theoretical expectations of an increasing use of local search operators with ongoing optimization progress.

One advantage of the adaptive control of the operators in each metaheuristic is the reduction of parameters that have to be chosen. The remaining parameters were calibrated by varying only one parameter while the others remained constant. A better parameter set could be obtained by conducting the calibration of all parameters simultaneously. Especially for the TA with four parameters, a more efficient search should be the result. In addition to the parameters, alternative decisions within each metaheuristic could be tested. For example, different threshold reduction schedules exist that could replace the schedule used. Or for the GA, the number of solutions to be replaced in one iteration could be increased. Another option is to change the termination criteria for each technique.

In this approach, only three different metaheuristics were tested, each representing a standard or simple variant of its type. Especially for GA, there is much research on the theory of metaheuristic optimization, which should be transfered into practical application. Thus, for this planning problem, the consideration of state-of-the-art theory regarding the search strategy or the operators and parameters might increase solution qualities and the search processes' efficiency. Another approach towards this direction would be to replace the search operators by operators that use problem-specific knowledge to lead the search towards good regions within the search space. For example, instead of randomly changing airports or departure times, remaining market sizes could be taken into account. Furthermore, additional metaheuristics might be easily applied. Instead of customizing one single search technique, in this study the basic design elements of metaheuristics are specified. Since these elements can constitute different (variants of) metaheuristics, the development of additional techniques is straightforward.

4.5 Evaluation

The objective of this section is to assess and compare the two different airline scheduling approaches presented in the previous two sections. For this purpose, each

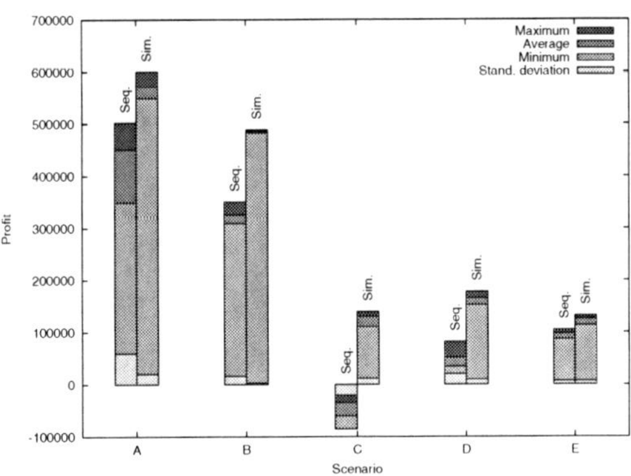

Fig. 4.66 Comparison of profits

calibrated approach is applied to the same set of planning scenarios. A comparison assesses the quality of the different solutions obtained by each solution technique and the efficiency of the search process. Then, using the method with the better performance, schedules are optimized for systematically modified planning scenarios to verify changes in the obtained solutions and the solution process.

4.5.1 Comparison

In sections 4.3.4.1 and 4.4.3.1, the best parameter setting for the sequential and simultaneous approach was determined (and the best search strategy for the metaheuristic, respectively). In addition, each calibrated model was applied to the five test scenarios (see Sect. B in the appendix). The results from these tests are presented in the following to compare both approaches.

Fig. 4.66 presents information on the profit[34] obtained by each solution approach for each scenario. On average, the simultaneous approach resulted in more profitable schedules for all scenarios than the sequential approach. Even when looking at the individual runs, the best schedule obtained by the sequential approach has a lower profit than the worst schedule from the simultaneous approach.

The higher profit of the schedules from the simultaneous approach might be explained by their higher seat load factors (see Fig. 4.67). As for the profit, not a single run of the sequential approach was able to obtain higher load factors than the worst run of the simultaneous approach.

However, as figures 4.68 and 4.69 show, the seat load factors result from different numbers of passengers and flights. For scenario B and C, the number of flights and the number of passengers were higher for the simultaneous approach than for the sequential approach. For scenarios A and E, the situation is vice versa, however, resulting in the lower seat load factor of the sequential approach. In scenario D, the

---

[34] Because of the high penalty costs in the simultaneous approach reducing the fitness value if maintenance restrictions are violated, every solution obtained fulfills this restriction. Thus, the fitness values of the solutions of the simultaneous approach match the profit.

Fig. 4.67 Comparison of seat load factors (SLF)

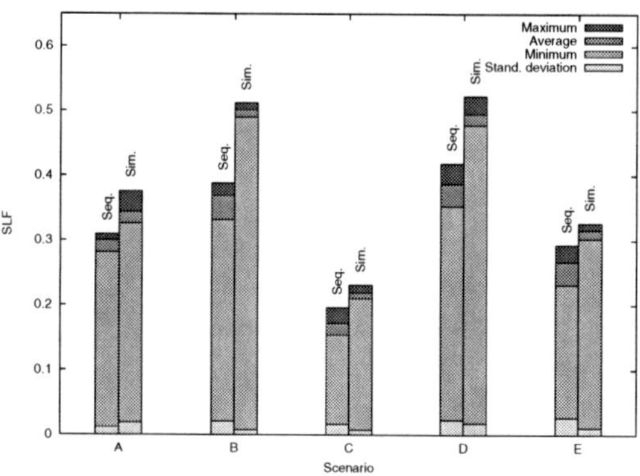

Fig. 4.68 Comparison of numbers of passengers

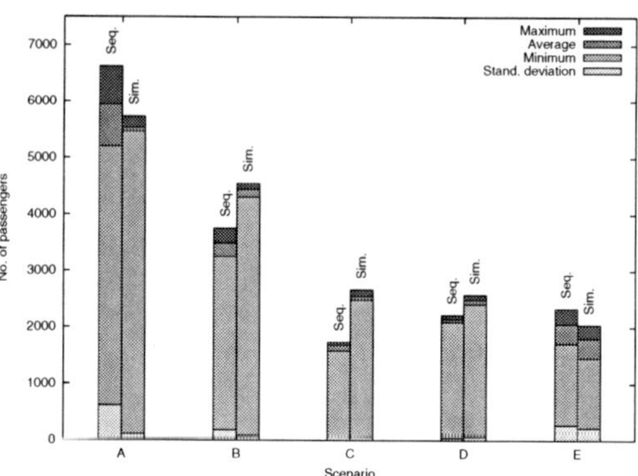

Fig. 4.69 Comparison of numbers of flights

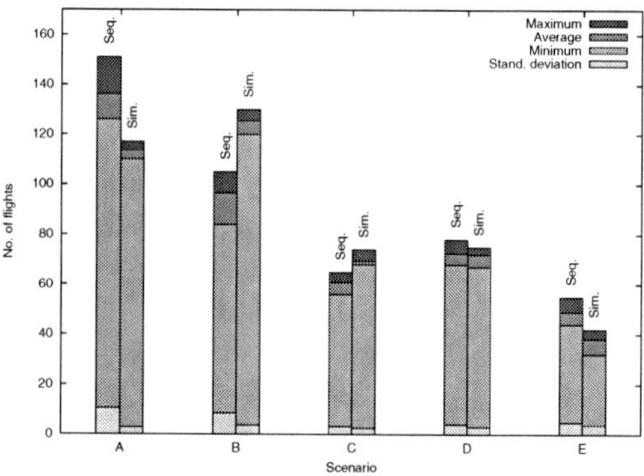

number of flights was lower for the simultaneous approach. These flights must be more attractive for passengers, because in contrast the total number of passengers is higher than for the sequential approach with its higher number of flights.

Another interesting result when comparing the sequential and simultaneous approach is their standard deviation of the key figures presented (see Table 4.7) on page 123 and Table 4.9 on page 151. In almost every experiment they were lower

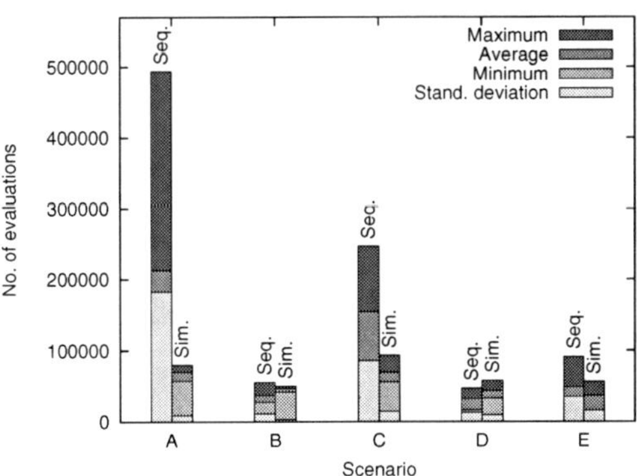

Fig. 4.70 Comparison of number of necessary fitness evaluations

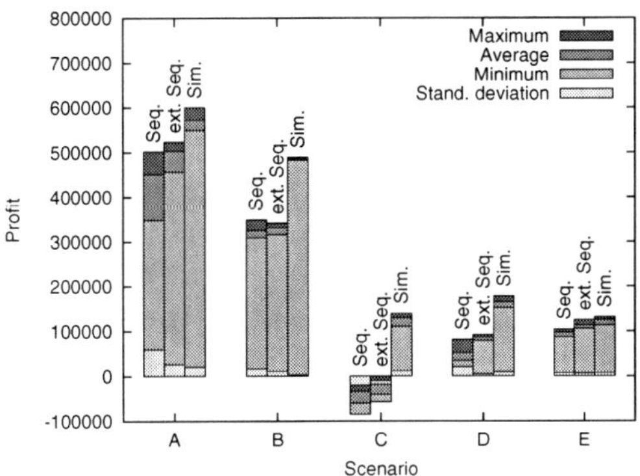

Fig. 4.71 Comparison of profits of schedules constructed with the sequential (regular and extended termination criteria) and simultaneous planning approach

for the simultaneous approach. Because the initialization of the GA is conducted randomly with every decision variable (airports and ground times) having the same selection probability, the low standard deviations of the schedules obtained are very encouraging. The – compared to the simultaneous approach – higher standard deviation of the sequential approach is most likely the result of the combination of deterministic and stochastic search during optimization. The initialization is conducted deterministically and the optimization steps perform a greedy – thus, rather deterministic – search, which should result in similar solutions in the end. However, because of the repair mechanisms and the maintenance routing steps, more significant and stochastic changes are applied to the solutions in the different runs. Because the deterministic search steps are then based on these modified solutions, the search processes take different paths through the search space, resulting in higher standard deviations of the final solutions.

In Fig. 4.70, the number of fitness evaluations that were necessary to obtain the final solution are compared. The GA needs considerably less effort to obtain its solution compared to the traditional approach. Especially for scenarios A and C the GA required only a fraction of evaluations compared to the number of evaluations the sequential approach needed. It has to be emphasized that when comparing both

Table 4.10 t-values for the validation of the solution approach comparison

Scenario	A	B	C	D	E
t-value	6.439	41.899	25.000	21.986	3.708

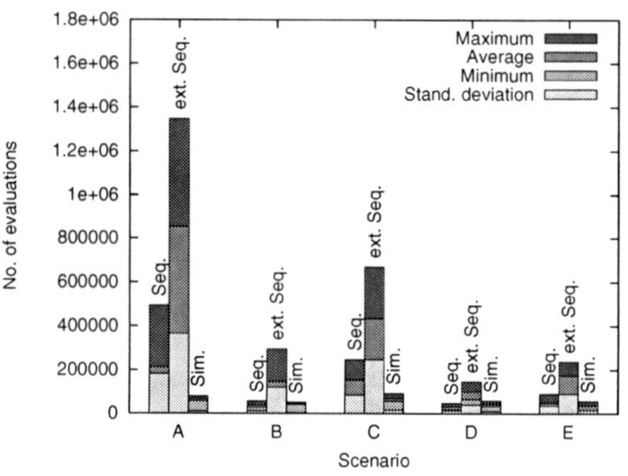

Fig. 4.72 Comparison of number of necessary fitness evaluations required by the sequential (regular and extended termination criteria) and simultaneous planning approach

solution approaches the number of iterations is an indicator for the computational effort and complexity. To compare the efficiency of the search in terms of improvement steps, the model specification has to be considered. For example, the sequential approach performs a greedy search in each optimization step, thus, many fitness evaluations are conducted before actually applying the optimization step (modification of the current schedule). In contrast, in the GA every fitness evaluation might result in an increase of solution quality – if the randomly applied modification was beneficial. However, these differences result from the underlying optimization techniques and are only of theoretical interest. The number of fitness evaluations necessary determines the effort of the techniques, resulting in the simultaneous approach being the technique that requires less computational effort and time.

To confirm the main results from the comparison – the simultaneous approach generates more profitable solutions with less effort than the sequential approach – a second optimization run is conducted for all scenarios using the sequential approach. To increase the probability of obtaining high quality solutions, its running time is further increased. For each scenario, the optimization run that required the most fitness evaluations is determined. Then, this number is doubled and used as termination criteria for the scenario. For example, the longest run for scenario A required a total of 698,906 fitness evaluations (see Table C.12 in the appendix). Thus, in the following experiments on scenario A the optimization is terminated if the number of fitness evaluations reaches $2 \cdot 698{,}906 \approx 1{,}400{,}000$, suspending i_{max} as termination criteria. For this second set of experiments with the extended termination criteria (ext. Seq.), results of the obtained schedule's profit and the number of fitness evaluations until the best solution was found are presented in the following figures 4.71 and 4.72. As a comparison, they also include the results from the first experiments of the sequential approach and from the simultaneous approach that were presented in figures 4.66 and 4.70.[35] The results clearly indicate that the GA-based simultaneous approach still outperforms the extended sequential approach, although it produced more profitable schedules compared to its first application (at the cost of significantly increased effort).

To validate the results, a standard Student's t-test is conducted for the results on the profit of both approaches.[36] The null hypothesis H_0 is that the observed differences in the profit values are random. H_α says that the differences are a result of the solution approach. The critical t-value for $p = 0.995$ is 3.335, for $p = 0.999$ it is 4.501. The results presented in Table 4.10 for the five scenarios show that the t-values always exceed the critical t-value of the level of significance. Thus, H_0 can be rejected on the 99.9%-level for all experiments except for scenario D, for which H_0 can be rejected on the 99.5%-level. As a consequence, the simultaneous approach significantly produces more profitable schedules than the sequential approach while requiring considerably less effort to obtain these schedules.

4.5.2 Experimental Verification

In the following, the simultaneous planning approach is used to solve airline scheduling problems of specific scenarios. By systematically modifying scenarios, the solution process and the obtained schedules are analyzed. These experiments also demonstrate the capability of the planning approach, representing a flexible decision support that is not limited to specific scenarios or assumptions regarding the structure of the market environment. Thus, it can be used for the assessment of different scenarios or variants and to assist in making strategic decisions (for example on the fleet sizes) in practice.

In each experiment, the calibrated GA is applied. Scenario D was chosen as a basic scenario, which then is systematically altered to analyze its effect on the solutions obtained. In the basic setting of scenario D, 50 airports can be selected for scheduling. A total of 20 aircraft of two fleet types is available: 10 Boeing 737-800, 10 Canadair Regional Jet 700.

4.5.2.1 Market Structure

The following experiments illustrate how changes in the market structure affect the schedule. A change in the market structure could for example be a shift of passenger demand to new cities or regions, or an abrupt drop in demand after external shocks or an economic downturn. To assess the capability of the solution approach to obtain appropriate solutions for such cases, the market sizes of the passenger demand are systematically changed so as to result in certain network structures of the airline schedules. In a first set of experiments, the market sizes are modified to result in a

---

[35] Individual results are presented in tables in Sect. C.3.

[36] The required test of the results for normal distribution was conducted using a Kolmogorov-Smirnov test.

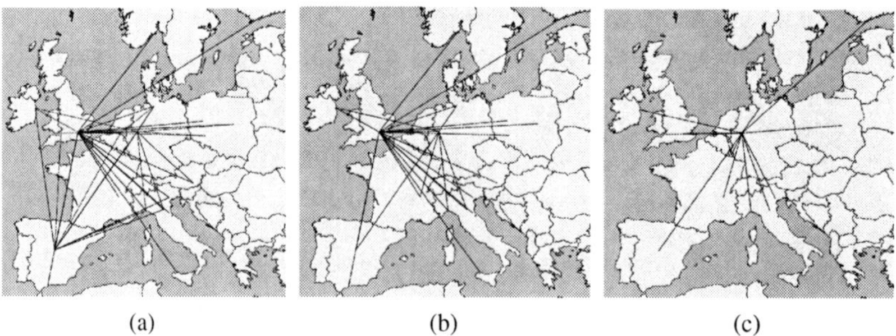

(a)   (b)   (c)

Fig. 4.73 Resulting hub-and-spoke network structure

hub-and-spoke network. Then, market sizes are chosen so that the resulting network should reflect a triangular-shaped route network (three routes).

Hub-and-Spoke Structure. Three experiments on three different market structures are conducted. In the first experiment (a), market sizes are zero for all city-pairs except for routes to or from Dortmund (DTM), London Heathrow (LHR), and Madrid Barajas (MAD). Thus, it should result in a hub-and-spoke network with DTM, LHR, and MAD as hubs. In the second experiment (b), the demand for routes to and from MAD is also set to zero (hub-and-spoke network with DTM and LHR as hubs); and in the third experiment (c) only DTM generates demand with all other city-pairs having no demand (single-hub network). The following figure 4.73 presents the network structures of the optimal schedules obtained for each experiment. The hub-and-spoke network structure is clearly visible for each experimental setup.

To illustrate the search process of the GA and its continuous modification of the network structure towards the final network, the following Fig. 4.74 presents networks that were processed during the optimization. As an example, the single-hub network is chosen. Based on one optimization run, the presented networks correspond to the best schedule in the population after initialization, 2% iterations (compared to the total number of evaluations needed for the complete run), 10%, 25%, and 50% iterations. Because the initialization of the GA run is conducted randomly with every decision variable having the same selection probability, the route network is spread over all available airports at the beginning. Then, the network continuously shifts towards the single-hub network with DTM as hub representing the optimal network structure for the given input data.

Routes. The following figures present the progress of the search process for the scenario in which market sizes between only three airports are given (DTM, LHR, MAD). Thus, a triangular-shaped route network should reflect the best network structure. In Fig. 4.75, the development from a randomly initialized route network towards the optimal solution is clearly visible.

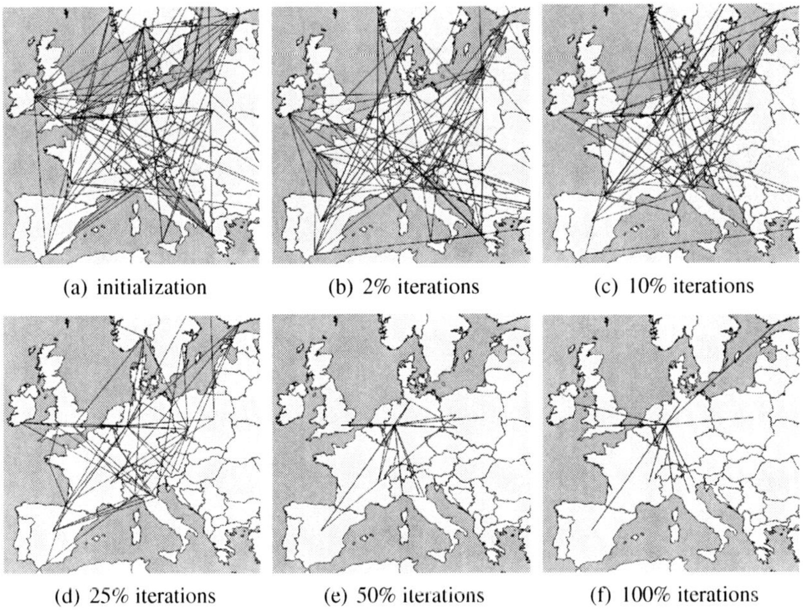

(a) initialization   (b) 2% iterations   (c) 10% iterations

(d) 25% iterations   (e) 50% iterations   (f) 100% iterations

Fig. 4.74 Illustration of the search process resulting in a single-hub network

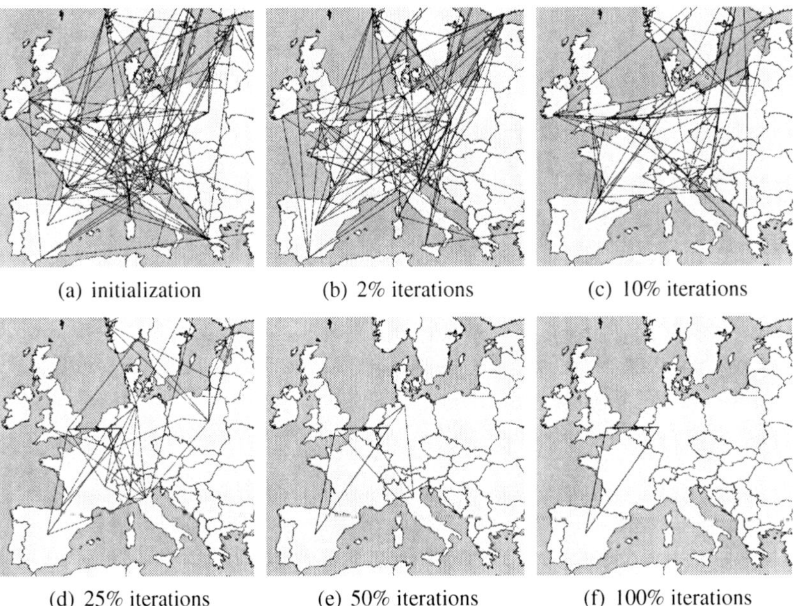

(a) initialization   (b) 2% iterations   (c) 10% iterations

(d) 25% iterations   (e) 50% iterations   (f) 100% iterations

Fig. 4.75 Illustration of the search process resulting in a network consisting of three routes

Fig. 4.76 Profit and number of required fitness evaluations for different numbers of aircraft

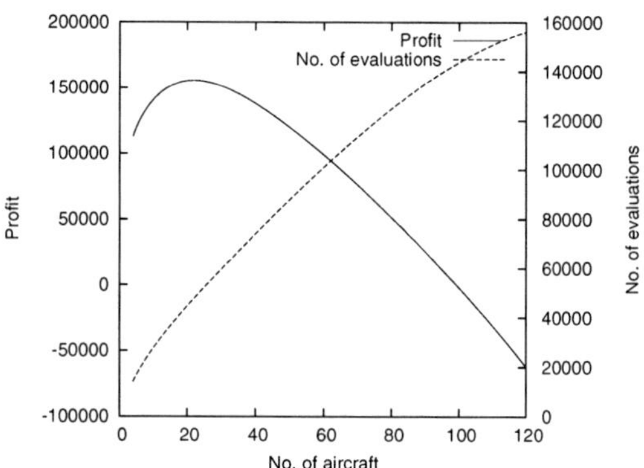

Fig. 4.77 Profit and number of required fitness evaluations for different sizes of the airport set

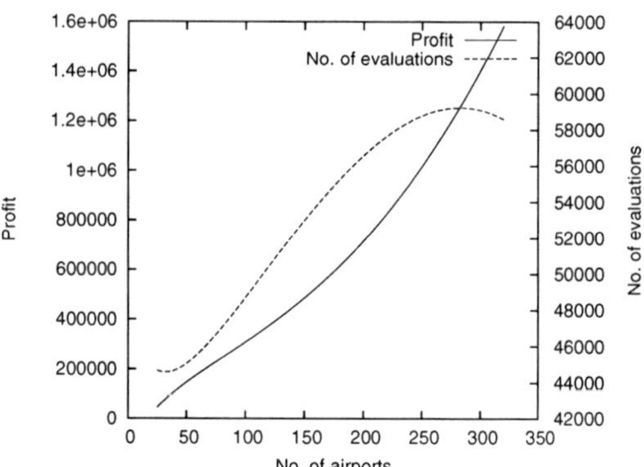

4.5.2.2 Number of Aircraft

The basic configuration of scenario D includes 20 aircraft of two different fleets of equal size. In the following, results of experiments are presented that use different numbers of aircraft. For example, if more aircraft are available, more passengers can be carried, possibly resulting in higher profits. On the other hand, if there are too many aircraft, their operational costs might exceed the revenues. Thus, finding the optimal fleet size for a given scenario is not easy and represents an important planning problem that can be supported by the simultaneous planning approach presented in this study.

The following Fig. 4.76 presents the operating profit and the number of required fitness evaluations for different fleet sizes. Following the basic configuration of scenario D, the distribution between both fleet types is kept at equal size. A peak in this figure can be identified, indicating the optimal fleet size. This number represents the best trade-off between too many aircraft causing high operational costs and too few aircraft reducing the number of passengers that can be transported.

4.5.2.3 Number of Airports

Every scenario considered in this study consists of a fleet composition and a set of airports. The solution approach then is allowed to choose only those airports that are included in this given set. This restriction should reflect the decision of some airlines to exclude certain airports from their schedules. In the following Fig. 4.77, results of experiments on different sizes of the airport set are presented. If the number of airports is increased, solution quality is higher, since there is more freedom in planning and more profitable flights can be selected. It is at its maximum if all 320 available airports can be selected for scheduling. However, the higher degree of freedom in search also requires more effort to obtain the final solution, thus, the number of fitness evaluations also grows with an increasing number of airports available for scheduling.

4.5.2.4 Fleet Types

In the following, an example is given on how different fleet types affect operating profit. Using the same number of aircraft, schedules are optimized with different fleet types. For clarity, in these experiments, the total fleet of the scenario only consists of the fleet type currently under investigation, thus, in each run 20 aircraft of the same fleet type are given. Besides the two types already used in the preceding experiments, four additional types from the 38 possible types (see Sect. A) are examined. They are selected with respect to differences in their operational characteristics (seat capacities). The total set of fleet types then consists of:[37]

- Fairchild Dornier 328JET (FRJ), seat capacity: 32
- Canadair Regional Jet 700 (CR7), seat capacity: 70
- Boeing 737-800 (738), seat capacity: 161
- Airbus A330-200 (332), seat capacity: 243
- McDonnell Douglas MD11 (M11), seat capacity: 279
- Boeing 747-400 (744), seat capacity: 373

Fig. 4.78 presents the profit for the different fleet types (seat capacities in parantheses). In general, for higher capacities of the fleet types, the profit decreases and becomes negative. As Fig. 4.79 illustrates, this is most likely the result of the reduced seat load factor. For the given demand in the scenario, these fleet types are oversized and the small number of passengers results in revenues too low to compensate the higher operating costs. These results correspond to reality, since airlines usually do not use large fleet types like the McDonnell Douglas MD11 or the Boeing 747-400 for flight service within Europe. The smallest aircraft type (FRJ) has the highest seat load factor, because it is easy to fill each aircraft with passengers. However, the profit of this fleet type is lower than for larger aircraft, indicating that the number of potential passengers is higher than the capacities offered.

---

[37] Fleet codes are in parentheses.

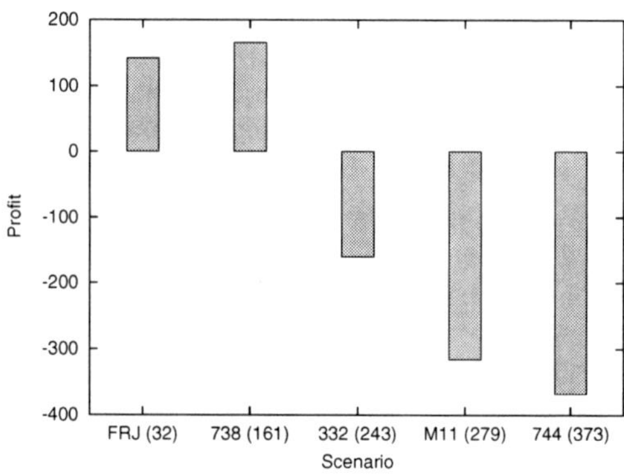

Fig. 4.78 Profit for different fleet types

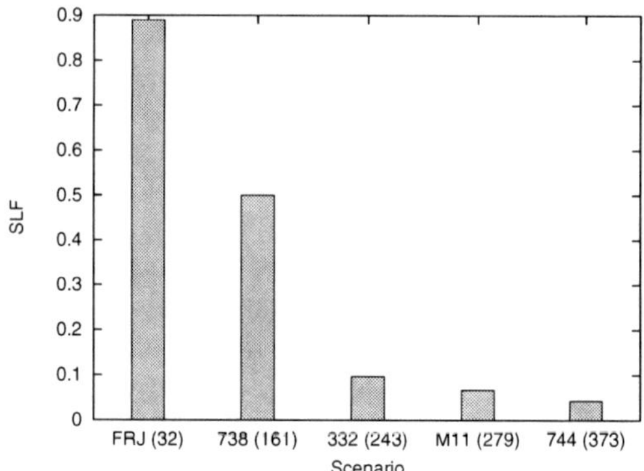

Fig. 4.79 Seat load factor (SLF) for different fleet types

4.5.3 Summary

This section consists of two parts: first, a comparison of the sequential and the simultaneous airline scheduling approach is conducted; then, the simultaneous approach as a more profitable technique is applied to modified scenarios to check the solution behavior and to demonstrate the capability of the planning approach.

The comparison of both approaches is conducted with respect to the operating profit of the solutions obtained and the required computational effort. For this purpose, both approaches are applied to five test scenarios. The simultaneous approach outperforms the sequential approach with respect to both criteria. Even when doubling the application duration of the sequential approach in order to increase solution quality, the GA-based planning approach yields better airline schedules while requiring only a fraction of the computational effort of the sequential approach. These results are confirmed by a standard Student's t-test on the schedules' profits.

Applying the simultaneous approach to different additional planning scenarios resulted in very satisfying observations. The scenarios represent systematic modifications of a basic scenario to investigate the solution behavior. Furthermore, they demonstrate the capability of the approach as a general decision support system that finds optimal schedules based on any given scenario. For example, if market sizes are modified so that demand only exists to and from one single airport, the GA is able to produce a final solution consisting of the optimal network structure: a single-hub network. Another example is the number of airports available for scheduling. If this number is increased, there is more freedom in scheduling and more profitable routes can be selected. The experiments confirm this hypothesis, their solution quality increased with an increasing number of airports and highest if all airports are available. Of course, the computational effort also increased due to more freedom in optimization. Finally, the experiments on the fleet composition give an example on how the approach can support (strategic) decisions on fleet sizes and types. For example, the impact on profit of different numbers of aircraft was tested and an optimal fleet size could be obtained for the given scenario. This number then represents the best trade-off between a number of aircraft that is too low and thus limits the number of passengers and too many aircraft causing high operational costs that the revenue from additional passengers cannot compensate.

4.6 Summary, Conclusion, Limitations, and Future Work

4.6.1 Summary

In this section, two approaches for integrated airline scheduling were presented and evaluated. They integrate the first two phases of airline scheduling with their sub-problems network design, frequency assignment, flight scheduling, fleet assignment, and aircraft routing. Furthermore, a schedule evaluation procedure was developed and calibrated that is required by both airline scheduling approaches. The development of this procedure was necessary because due to their proprietary nature commercial applications or the required data and parameter sets were not available for this study.

Besides integrating the airline scheduling problem, both planning approaches are able to represent airline operations and practical requirements on a higher level of detail compared to many solution models presented so far. In addition, there are fewer simplifying assumptions or restrictions to certain planning scenarios (for example uniformly distributed demand, monopoly airline, only one fleet type, (single) hub-and-spoke network etc.).

The first approach follows the traditional planning paradigm of decomposing the overall problem into less complex subproblems that are solved in a sequence. This stepwise approach is realized in an iterative procedure consisting of solution models from literature. To assist each planning step in finding a feasible solution and to improve the applicability of the complete planning procedure, many supportive functions had to be implemented. As a result, this procedure is rather complex. In contrast, the second planning approach represents a truly simultaneous model. In a metaheuristic, each processed solution represents a complete airline schedule, thus including all former subproblems implicitly. An adaptive search process was developed that controls the application of the search operators based on their past profit contribution. All parameters of both models were calibrated using (European) real-world data to lead to the best solution quality. For the metaheuristic approach, a genetic algorithm (GA) was identified as the most effective search strategy.

An analysis of the solution quality and search process of the calibrated models produced satisfactory results. The results were very stable, thus, although both models inherit heuristic elements, the resulting schedules of different optimization runs for given planning scenarios yielded similar schedules. However, differences exist in the progress of the search process. While the GA-based planning approach shows a continuous improvement of solution quality that corresponds to theoretical expectations (also with regard to the adaptive control of the search operators), the optimization progress of the sequential planning approach is characterized by a very unstable trend. Peaks and drops in profit can be observed even for succeeding iterations. A closer examination unveils that this effect is most likely the result of the sequential planning paradigm with its insufficient consideration of interdependencies between the subproblems. As a consequence, for example, the maintenance routing has to apply many changes to a schedule given by the previous fleet assignment to make it feasible with regard to maintenance restrictions. Since the objective of these modifications is to create feasibility without considering economic implications, the drops in profit during the total search process can be easily explained. Thus, in future work, a better integration of the fleet assignment and the maintenance routing should be the first starting point when improving the sequential planning approach presented in this study.

A comparison of both airline scheduling approaches is straightforward since they use the same set of input data. Applying both approaches to the same scenarios results in the GA-based simultaneous approach being the more efficient planning technique. Verified by a Student's t-test, this approach resulted in more profitable schedules for all scenarios while using only a fraction of the computational effort compared to the sequential approach. The capability of the simultaneous planning approach is further investigated by its application to scenarios that were modified implying a certain structure of the optimal solutions. For all experiments, the resulting schedules are in accordance with theoretical expectations. For example, if market sizes are set to zero except for routes originating or departing at one specific airport, a schedule with a single-hub network was obtained as a final solution. These kinds of experiments also give an example on how the approach can assist in making (strategic) decisions in airline planing. Only the given planning scenario consisting of available airports, the number of aircraft, and the fleet composition has to be provided to the integrated approach; then, an airline schedule that best fits to the given scenario is automatically constructed and optimized.

4.6.2 Conclusion

In general, researchers agree that integrating the subproblems of the airline scheduling problem in one model should result in a higher solution quality. The comparison in this study of the sequential and the simultaneous airline scheduling approach supports this statement, since the simultaneous approach constantly yields higher operating profit of the optimized schedules. The general drawback of the sequential approach is the non-consideration of interdependencies between the individual subproblems. In each solution step, a different optimization problem is solved, each with its own objective function. These objectives might be contradictory and not congruent with the overall objective. As a consequence, the search for the optimal schedule is not straightforward but characterized by many changes in the direction of the search, limiting its overall success. In contrast, in the simultaneous approach there is a permanent orientation towards the overall objective and every search step is conducted with respect to this goal.

Both approaches use the operating profit as optimization criterion. In the sequential approach, the optimization steps are conducted using a greedy search based on the schedules' profits. In the metaheuristic simultaneous approach, the operating profit corresponds to the fitness values of the solutions processed. Thus, instead of modeling a functional relationship between the decision variables and the objective value, existing solutions have to be assessed (however, this still represents a complex task). This allows the evaluation of schedules on a high level of detail and the integration of additional quality features without necessarily having to know their exact relationship to the decision variables. For example, factors like demand distributions over the day or competition not limited to a single airline are considered in the approaches presented here that have received little attention in past publications. In addition, the schedules can be assessed with regard to any optimization goal or additional (for example operational or managerial) restrictions can be taken into account as penalty costs. The optimization routines do not need to be changed for both approaches. As a result, a very flexible scheduling can be conducted. However, this flexibility shows its strengths best when using the simultaneous approach, because – as explained in the previous paragraph – then the entire search is focused on the given overall objective.

4.6.3 Limitations

Before presenting some possible directions for future work to overcome specific limitations of the presented airline scheduling approaches, one general characteristic of this study has to be emphasized. The findings of this study do not allow a universally valid conclusion or an evidence of the superiority of the simultaneous airline scheduling approach, because the sequential approach used as a benchmark represents only one possible way to construct an airline schedule with the sequential planning paradigm. It does not represent an exact reproduction of the status-quo of airline scheduling, since such a standard procedure that all airlines use or researchers accept simply does not exist. As already addressed, because of the high complexity of the problem there are multiple ways to approach the airline scheduling problem. Each airline has its own course of action to construct an airline schedule with its individual planning steps and responsibilities of experts. In addition, as was presented in Sect. 2, multiple approaches were developed in airline scheduling research, leading to a multitude of explicit procedures to construct an airline schedule.

To further validate the advantageousness of the simultaneous airline scheduling approach, additional benchmark tests with airline scheduling models using the sequential planning paradigm have to be conducted. However, until now models integrating the same subproblems as the simultaneous planning approach have not been developed or published. Thus, much effort is still necessary to conduct additional tests, especially when dealing with practices of human experts and within airlines.

4.6.4 Future Work

Future work should focus on two directions: minimizing some limitations of the airline scheduling approaches and their experiments presented in this study to further verify the essential findings and extending and improving the integrated planning approach for additional future applicability.

Some drawbacks and possible future enhancements of the schedule evaluation procedure and the airline scheduling approaches were discussed in the corresponding sections. To summarize, the basic restriction in schedule evaluation is the availability of data. If more data on a detailed level was available, it could replace the estimates of the airline schedule evaluation procedure or be used to calibrate the presented steps on a higher level of confidence. Then, for example, the estimation models and their parameters can be different among markets, time periods, and passenger segments (leisure and business). For the sequential scheduling approach, especially the maintenance routing step has to be more smoothly integrated into the overall planning, since this step is responsible for the unstable search progress and the resulting lower solution quality. The conceptual design of the simultaneous approach represents general design elements of a metaheuristic not limited to the search strategies used. Thus, testing additional metaheuristic search concepts and applying more sophisticated versions of the used techniques can be conducted to further increase solution quality.

The flexibility of the solution approach described in the previous section enables many starting points for further enhancements. Additional operational or managerial objectives could be included in optimization. For example, factors like regularity of the flights, the length of rotations per fleet, or economies of scale at hub airports can be easily included in the objective function. A desirable feature might also be to favor solutions that contain a large number of aircraft on the ground at the same time to allow modifications of the rotations in the operational planning. If airport slots need to be considered, penalty costs can be adjusted so that they represent the expenses or efforts to acquire a slot. Penalty costs can also be used if an airline wants to include specific flights in a schedule because of marketing or strategic decisions (and these flights do not produce an innate profit). If an airline wants to construct an airline schedule based on an existing schedule, (slightly modified) copies of the old schedule can be included in the initial population if using the GA-based solution approach or the old schedule is used as an initial solution in the other planning approaches. If only a modified version of an old schedule should be generated, deviations between the old schedule and the new schedule result in penalty costs. Thus, changes to flights are only allowed if their additional profit outweighs the changes in the schedule.

One important element in airline scheduling that was excluded from the solution approaches presented is crew scheduling. However, given an airline schedule, some crew costs are fixed, too. Thus, when considering these crew costs in the fitness function, better overall solutions can be achieved. For example, if an aircraft has to overnight at an airport that is not a crew base, the additional costs for accommodation expenses (or for repositioning flights) for the crew members can be considered as penalty costs. In addition, if the aircraft is scheduled to depart before the minimum overnight crew rest has elapsed, penalty costs for the lonely overnight of the crew are assigned.

Finally, the integrated airline scheduling is not limited to passenger airlines but can also be used for cargo airlines that offer a scheduled service. However, most freight is transported by combination carriers which carry cargo and passengers on scheduled passenger flights (Doganis, 2004). For these airlines, Link (2006) presents a model for a fair allocation of costs and revenues among passengers and cargo and an efficient metaheuristic for multi dimensional optimization of package flows. Because – compared to algorithms based on standard models for multi commodity flow problem – its solution time is significantly low, it could be integrated in the objective function of the airline scheduling approaches. As a result, airline schedules are optimized to represent the best trade-off of revenues gained by passenger and freight transportation.

Chapter 5
Summary, Conclusions, and Future Work

Abstract. In this study, two airline scheduling approaches were developed that integrate the flight schedule generation and aircraft scheduling phase into a single scheduling approach. One of the two approaches for airline schedule optimization follows the traditional planning paradigm of iteratively and sequentially solving subproblems of the overall airline scheduling problem. The other airline scheduling approach is based on self-adaptive metaheuristic optimization in which complete airline schedules are processed at once. Applying both approaches to the same scenarios results in the simultaneous approach being the more efficient planning technique. The capability of the simultaneous approach is further demonstrated by verifying its results for systematically modified planning scenarios. The simultaneous planning approach of this study optimizes a large portion of the overall airline scheduling problem in an integrated procedure while minimizing simplifying assumptions. Thus, many of the requirements formulated in airline operations research literature are fulfilled. However, further challenges exist that future work should focus on: incorporating the complete crew planning into this scheduling approach, including stochastic elements in the schedule evaluation to minimize the effects of disruptions, further increasing the level of detail in which airline operations are represented and considerung more practical requirements, and finally – since this study represents a theoretic framework – assessing the applicability of the integrated approach in real-world airline scheduling.

5.1 Summary

Since its beginning as exclusive adventure airline travel has become a mass travel system representing one of the most valuable assets for economic growth. In the past, a constant increase of the total passenger kilometers of scheduled passenger airlines could be observed that is expected to continue in future years. However, despite this positive trend, the airlines profit margins are considerably small and strongly depend on overall passenger demand. As a result, the airlines' profitability is cyclical, following economic upturns and downturns. For each airline, the challenge is to match its resources like personnel and aircraft to the demand given by the market. The instrument to accomplish this task is the airline's schedule, containing all flights of

T. Grosche: Computational Intel. in Integrated Airline Scheduling, SCI 173, pp. 173–176.
springerlink.com

the airline and the assignment of the resources. Hence, an optimal schedule represents the most efficient and effective deployment of an airline's resources while best satisfying potential passenger demand. It is the central element within an airline's corporate planning system, because it affects almost every operational decision and has the largest impact on profitability.

As a consequence, the construction of an airline schedule is one of the most important but also most complex planning tasks of each airline. Many factors such as demands in various markets, competition, and available resources have to be taken into account. Unfortunately, a single optimization model for the complete airline scheduling problem is intractable when using exact optimization techniques. Instead, this problem is solved in a sequential approach. The overall problem is decomposed into subproblems of less complexity; these subproblems are solved in a sequence, and the solution of one problem serves as input for the next problem. Some subproblems are grouped together to form airline scheduling phases. One possible decomposition of the overall problem and aggregation of the subproblem to scheduling phases is proposed on page 10. Many different solution approaches were developed for individual planning steps. An extensive presentation of these models and the underlying problems including their objectives and constraints are given in Sect. 2. Since in general a decomposition of a problem cuts interdependencies between decision variables, and a solution sequence limits flexibility of later planning steps, only suboptimal or even infeasible solutions of the problem can be achieved. To reduce these disadvantages for the airline scheduling problem, airlines usually implement iterations in the planning process where solutions or details of later planning steps are processed to earlier steps. However, since it is impossible that a sequential solution approach can achieve better or equal solutions than a simultaneous approach, research focuses on the integration of different subproblems into a single optimization model. Models that aim at integrating selected subproblems are presented in Sect. 2.5.

The objective of this study is to fill a large gap between the status quo in airline scheduling and the optimal scheduling using a fully integrated optimization model that includes all subproblems and represents airline operations on a sufficient level of detail. For this purpose, two airline scheduling approaches were developed that integrate the flight schedule generation and aircraft scheduling phase into a single scheduling approach. Their only requirement is to receive a quality measure for each schedule processed. As schedule evaluation applications used by airlines and their required parameters and data are not available for this study, a custom evaluation procedure was developed that estimates the operating profit for any given airline schedule (Sect. 4.2). One of the two approaches for airline schedule optimization (presented in Sect. 4.3) follows the traditional planning paradigm of iteratively and sequentially solving subproblems of the overall airline scheduling problem. For the individual solution steps, existing models from literature were used, which are then integrated in a complete planning procedure. The other airline scheduling approach (presented in Sect. 4.4) is based on self-adaptive metaheuristic optimization in which complete airline schedules are processed at once. Because in each schedule the subproblems and interdependencies are included implicitly, the optimization results in a truly simultaneous airline scheduling approach.

Ａ comparison in which both approaches are applied to the same scenarios confirmed the postulated higher performance of a simultaneous optimization since the simultaneous approach outperformed the sequential approach with regard to the operating profit of the obtained schedules and the required computational effort (Sect. 4.5.1). The capability of the simultaneous approach is further demonstrated by verifying its results for systematically modified planning scenarios (Sect. 4.5.2).

5.2 Conclusion

The simultaneous planning approach of this study optimizes a large portion of the overall airline scheduling problem in an integrated procedure while minimizing simplifying assumptions in comparison to existing solution models. It can be used for decision support for flexible airline scheduling, because it only requires given external data and the supply data from an airline. Furthermore, the objective of scheduling is not limited to maximizing operating profit but can include any quantifiable goal.

Thus, many of the challenges or requirements formulated in state-of-the-art airline operations research literature are fulfilled. The main objective – further integrating subproblems towards the ideal model of a fully integrated overall scheduling approach – is achieved. Until now, an integrated model including the subproblems network design, frequency assignment, flight scheduling, fleet assignment, and aircraft routing has not been developed. All schedule elements that are assumed to be given in other approaches (like the network structure, number of hubs, etc.) are a result of optimization. Thus, this model represents the most integrative airline scheduling approach at this time. Experiments on the simultaneous and sequential approach were conducted that verify the postulated better performance of a simultaneous optimization for the test scenarios used in this study.

Compared to existing models, the planning approach of this study represents airline operations on a high level of detail without simplifying assumptions. For example, existing models assume uniformly distributed passenger demand, a monopoly situation, a single fleet, a given and static hub-and-spoke network structure etc. In contrast, the simultaneous planning approach presented here optimizes airline schedules for any given planning scenario. This allows a very flexible scheduling, since only given external data and the supply data of the airline have to be provided; a modification of the solution approach is not necessary. In addition, the ability to easily change the objective function or to include restrictions or managerial constraints as penalty costs further increases the flexibility of the approach. Changes in the input can be easily evaluated according to the given objective and operational impacts. Furthermore, an airline can apply *what-if* scenarios to review future directions and to test different courses of action. Thus, the planning approach enables a powerful decision support for airline scheduling.

5.3 Future Work

As described in the corresponding sections, many further enhancements to improve scope and solution quality are possible. The simultaneous solution approach provides large flexibility and allows easy modifications of the optimization objective or general conditions. Sect. 4.6.4 presents how some basic elements of crew scheduling can be included in the planning approach of this study. However, since crew costs represent one of the highest expenses of an airline, additional effort is necessary to incorporate the complete crew planning into this scheduling approach. If successfully accomplished, the resulting model should be close to the ideal model airline operations research demands, since all subproblems currently tackled independently could then be solved in one step.

Another challenge receiving much attention by researchers today is to increase the robustness of airline schedules. Traditional solution models are based on deterministic data, although many influencing factors are of stochastic nature. Thus, often the schedule is not executed as planned. For example, adverse weather or maintenance issues cause irregular operations in the scheduled activities. To minimize the effect of these disruptions, stochastic elements can be included in the schedule evaluation. Then for example, a schedule is not only evaluated according to the operating profit but also to the probability and the extent of possible delays caused by disruptions. Although such an assessment of a schedule might represent a complex task itself, it could be easily included as fitness function for the presented metaheuristic search.

Although the level of detail in which airline operations are represented is much higher in this study than in previous contributions, there is still much room for further enhancements regarding practical considerations. For example, the schedule evaluation procedure does not yet distinguish between business and leisure travelers and different seating configurations of the aircraft. Furthermore, curfew restrictions only take required runway lengths and a single period of night-flying restrictions per airport into consideration. In reality, these influences consist of many different elements that should be modeled in the airline scheduling approaches. Another limiting factor is airport slots. In Europe, the major airports usually have fewer slots available than airlines demand. Taking into account the expected future growth in airline traffic, slots will even more reduce the degree of freedom in airline scheduling. As a consequence, the scheduling procedures presented in this study should be further extended to include slot restrictions.

Until now, this study represents a theoretic framework; its applicability in real-world airline scheduling still has to be assessed. Hence, the planning scenarios and all input used in this study should be replaced by existing data from an airline. This also should enable the use of planning scenarios from regions other than Europe, to which this study was limited because of the availability of data. If possible and applicable, using the same scenario and prerequisites for optimization which real airline schedules were based on, the presented approach can be further evaluated and compared to the corresponding real-world schedules. Additional enhancements based on such practical experience would then advance the presented approach for integrated airline scheduling to an important and valuable optimization technique for both theory and practice.

Appendix A
Aircraft Data

The following table presents the aircraft-related information that is used in all experiments.

Fleet Name	Manufacturer	IATA Code	Capacity (Seats)	Block Hour Costs (US-Dollar)	Turn Times (min)	Range (km)	Required Runway Length (m)
328JET	Fairchild Dornier	FRJ	32	1,715	31	2.593	1,380
717	Boeing	717	113	3,776	44	3,180	2,131
737-200	Boeing	732	107	5,335	43	3,890	2,315
737-300	Boeing	733	128	3,362	45	3,612	2,530
737-400	Boeing	734	142	3,030	49	3,818	2,269
737-500	Boeing	735	106	3,034	45	2,658	2,496
737-700	Boeing	73W	44	2,673	33	3,585	2,393
737-800	Boeing	738	161	3,337	52	3,585	2,589
747-400	Boeing	744	373	8,434	87	12,846	3,027
757-200	Boeing	752	182	4,435	55	4,906	2,169
767-200	Boeing	762	199	4,439	58	6,495	2,325
767-300	Boeing	763	225	5,271	63	7,835	2,542
777	Boeing	777	313	6,440	105	12,011	2,706
A300	Airbus	AB3	260	5,480	68	3,430	2,264
A300-600	Airbus	AB6	256	8,050	65	5,285	2,284
A319	Airbus	319	116	2,555	46	3,396	2,097
A320-100/200	Airbus	320	149	2,532	51	4,022	2,454
A321-100/200	Airbus	321	183	2,636	53	4,276	2,256
A330-200	Airbus	332	243	5,928	66	11,675	2,448
ATR 42-300 / 320	Avions de Transport Régional	AT4	46	2,554	33	4,480	1,088
ATR 72	Avions de Transport Régional	AT7	65	2,851	36	2,665	2,013
BAe 146-300	British Aerospace	143	98	3,361	42	2,400	1,616
Canadair Regional Jet 100	Bombardier Aerospace	CR1	49	1,421	34	1,833	1,818
Canadair Regional Jet 700	Bombardier Aerospace	CR7	70	1,919	37	2,776	1,968
DC-9	Douglas	DC9	106	3,633	43	2,880	2,330
DHC-8-100	Bombardier Aerospace	DH1	37	1,269	32	1,780	1,115
DHC-8-400	Bombardier Aerospace	DH4	67	2,132	36	2,400	1,403
EMB 120	Embraer	EM2	30	1,235	30	1,224	1,563
Embraer 170	Embraer	E70	71	2,223	37	3,334	1,955
ERJ 135	Embraer	ER3	37	1,454	32	2,650	2,158
ERJ 145 Amazon	Embraer	ER4	49	1,650	34	2,648	2,091
F100	Fokker	100	100	2,477	42	2,505	1,824
Jetstream 41	British Aerospace	J41	29	1,243	30	1,433	2,039
MD11	McDonnell Douglas	M11	279	7,757	81	12,817	2,898
MD80	McDonnell Douglas	M80	146	3,507	38	2,897	2,619
MD90	McDonnell Douglas	M90	149	3,190	40	3,862	1,876
RJ85	Avro International Aerospace	AR8	93	1,777	41	2.400	1,617
SF340A/B	Saab Fairchild	SF3	33	1,396	30	2,387	1,663

T. Grosche: Computational Intel. in Integrated Airline Scheduling, SCI 173, p. 177.
springerlink.com   © Springer-Verlag Berlin Heidelberg 2009

Appendix B
Experimental Setups

Each scenario consists of airline-independent general data and the specific situation of the airline. In this study, the data described in Sect. 4.1.3 and in the Appendix A is used as general data, whereas the situation at the airline focuses on the following two elements:

- number of aircraft available, fleet composition, and maintenance stations,
- set of airports or markets that the airline is willing to accept in its schedule.

For each scenario, these elements are chosen according to the tables in the next section, which are selected to represent different possible scheduling problems. In addition, because the objective is to construct a daily airline schedule, competing flights from a random chosen day are included in the schedule evaluation process.[1] The following Table B.1 presents an overview of the five different scenarios including the day chosen for competing flights, the number of aircraft and fleets, and the number of airports available to the optimization process.

Table B.1 Test scenarios

Scenario	Day (in 2004)	No. of Aircraft	No. of Fleets	No. of Airports
A	March 19	30	4	62
B	August 3	30	3	29
C	July 27	28	4	55
D	February 9	20	2	50
E	June 5	10	1	90

---

[1] The schedule evaluation represents the bottleneck in terms of computation time. The required time depends on the number of flights and itineraries that have to be evaluated which in turn depends on the number of competing flights. To be able to conduct a sufficient number of experiments, the number of competing flights given by the OAG schedules is randomly reduced to 10% of its original value. This reduction does not bias the fundamental results, because this reduction is applied to all experiments and - to keep a realistic estimation of passenger demand - the given market sizes are also reduced to 10% of their original value.

T. Grosche: Computational Intel. in Integrated Airline Scheduling, SCI 173, pp. 179–181.
springerlink.com © Springer-Verlag Berlin Heidelberg 2009

B.1 Scenario A

Aircraft. 5 Airbus A319, 10 Airbus A320-100/200, 5 Airbus A321-100/200, 10 Boeing 737-800

Airports. AAL, AJA, AMS, ATH, BCN, BES, BMA, BOD, BRN, BRU, BSL, CAG, CDG, CGN, CPH, DRS, DUB, ESB, FCO, FLR, FRA, GOT, HAM, HEL, INN, IST, LEJ, LHR, LIS, LNZ, LUX, LYS, MAD, MAN, MMX, MUC, NAP, NCE, NRK, NTE, NUE, OPO, ORK, OSL, OTP, PMO, PRG, RIX, SOF, SPU, STR, SXB, SZG, THF, TKU, TLL, TLS, VCE, VNO, WAW, ZAG, ZRH

B.2 Scenario B

Aircraft. 10 Airbus A320-100/200, 10 Boeing 757-200, 10 Canadair Regional Jet 700

Airports. AMS, ARN, ATH, BRN, BRU, BUD, CDG, CIA, CPH, DUB, HEL, IST, KEF, LHR, LIS, LJU, LUX, MAD, OSL, OTP, PRG, RIX, SOF, TLL, TXL, VIE, VNO, WAW, ZAG

B.3 Scenario C

Aircraft. 8 Airbus A321-100/200, 8 Boeing 737-300, 4 Airbus A300-600, 8 McDonnell Douglas MD80

Airports. ACE, ALF, ASR, ATH, BCN, BHX, BIQ, BSH, BTS, CFR, CFU, DUB, DUS, EFL, FRA, FSC, GCI, GDN, GNB, GVA, GWY, HUY, JTR, KEL, KLU, LCJ, LCY, LGW, LPA, LPI, LPL, LUG, LYS, MJV, MMX, OLB, ORM, OSL, OUL, OXF, PMI, RJL, SOF, SPU, STR, SUF, TFN, UME, VAR, VCE, VIE, VIT, ZAD, ZAG, ZAZ

B.4 Scenario D

Aircraft. 10 Boeing 737-800, 10 Canadair Regional Jet 700

Airports. ADB, AGH, AGP, ALF, ANR, AVN, BGO, BIQ, BLQ, BOD, BRS, DTM, DUB, EIN, ERZ, FRL, FSC, GOA, GRZ, GWY, HAM, HAU, HEL, INV, JKH, KLU, KTT, LCG, LHR, LUG, LUX, LYS, MAD, NRK, NRN, OMR, ORK, PLQ, PMO, POZ, PVK, TLL, TRF, TSF, TZX, UIP, UME, WAW, WRO, ZAD

B.5   Scenario E

Aircraft. 10 Airbus A320-100/200 **Airports.** AAL, AAR, ACE, ADA, AGH, AHO,

ALF, ANR, AOK, AVN, AXD, BCN, BES, BGO, BIA, BIQ, BLE, BRN, BUD,
CDG, CFR, CPH, CWL, DNZ, DUB, DUS, EIN, ERC, ETZ, EXT, FLR, FMO,
FRL, FUE, GOA, GRZ, HAJ, HDB, HER, IAS, JCA, JKG, JOE, JYV, KGS, KID,
KOK, KRK, KRS, KUO, LBA, LCY, LDY, LEI, LGW, LIL, LIS, LLA, LPI, MJT,
MLA, MLX, MME, MUC, MXP, NRN, NUE, OPO, OTP, OVD, PAD, PLH, PLQ,
PMI, PNA, POZ, PVK, SKG, STN, TFS, TKU, TLN, TRD, TUF, VIT, VLL, WAW,
WRO, ZRH, ZTH

Appendix C
Experimental Results

In this section, results of all experiments conducted for this reserach are presented in detail. Whereas Sects. 4.3 and 4.4 primarily contain aggregate figures of the five scenarios, in the following results and trends of individual calibration and optimization runs are presented.

The following Sect. C.1 contains the calibration results of the parameters of the sequential airline scheduling approach (Sect. C.1.1) and all three search strategies (TA, rGA, and GA) of the simultaneous airline scheduling approach (Sect. C.1.2). Individual results when applying the calibrated models to decide about the search strategy of the simultaneous approach are presented in Sect. C.1.2.4. Then, Sect. C.2 focuses on the analysis of the sequential and simultaneous approach. Key figures of the solution quality of the simultaneous approach are presented in Sect. C.2.1.1, details of its solution process for the five scenarios in Sect. C.2.1.2. Similar results are presented for the simultaneous approach in Sects. C.2.2.1 and C.2.2.2. Finally, Sect. C.3 contains the key figures of the extended sequential approach.

T. Grosche: Computational Intel. in Integrated Airline Scheduling, SCI 173, pp. 183–229.
springerlink.com

C.1   Calibration

C.1.1   *Sequential Approach*

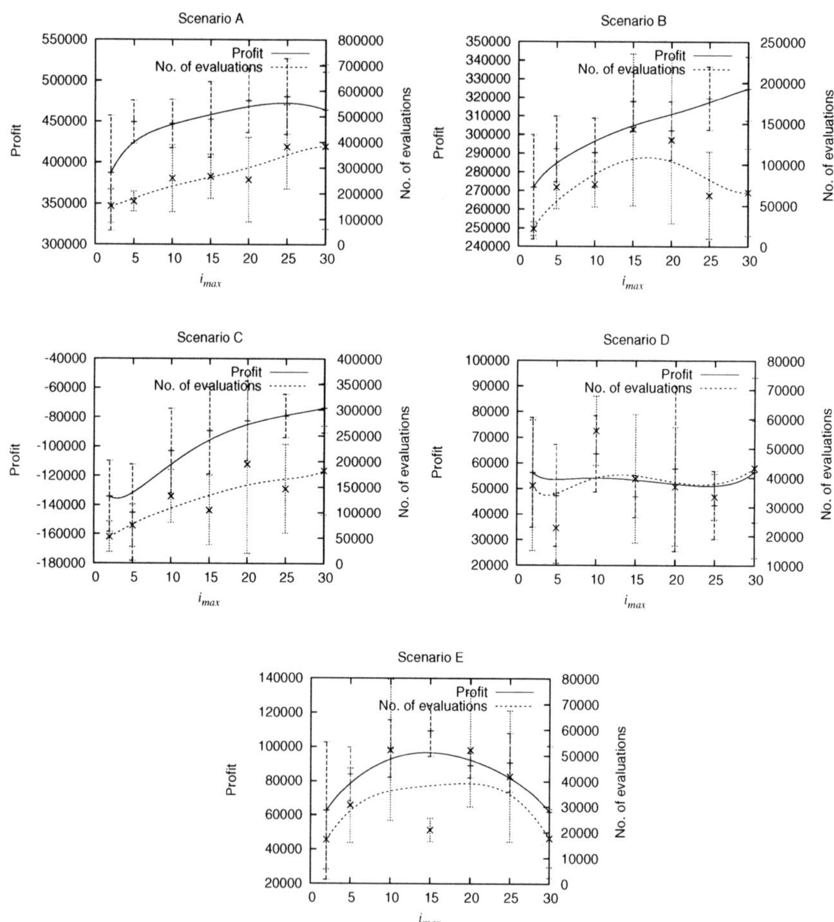

Fig. C.1 Calibration results for parameter i_{max}

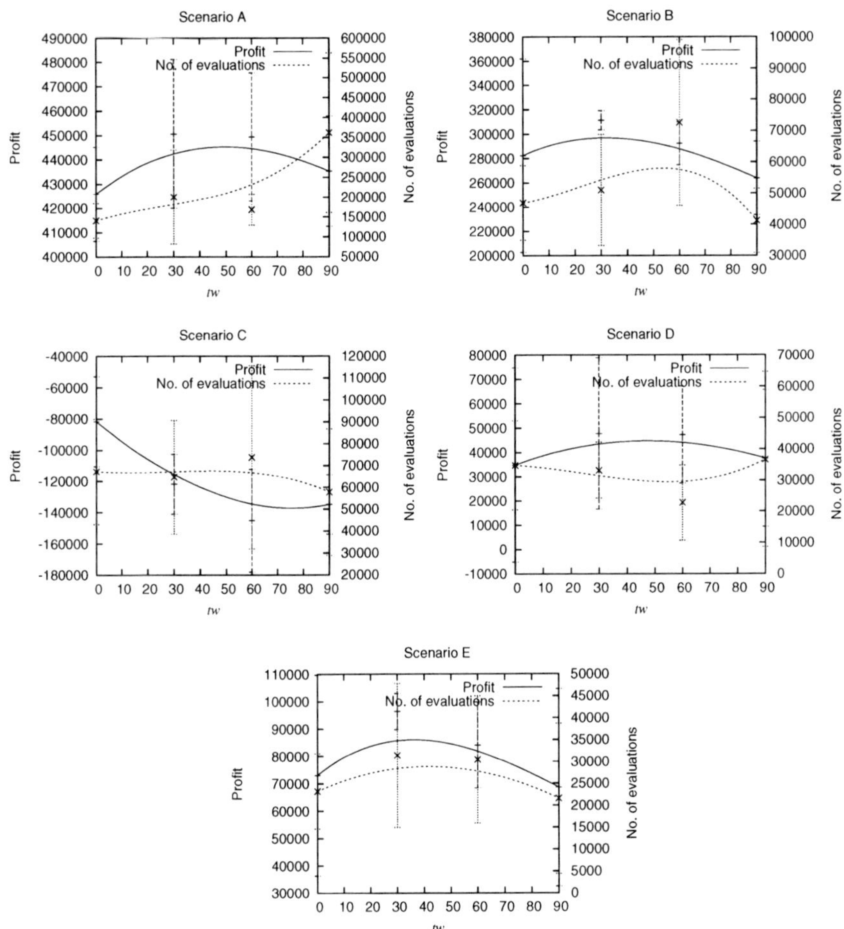

Fig. C.2 Calibration results for parameter tw

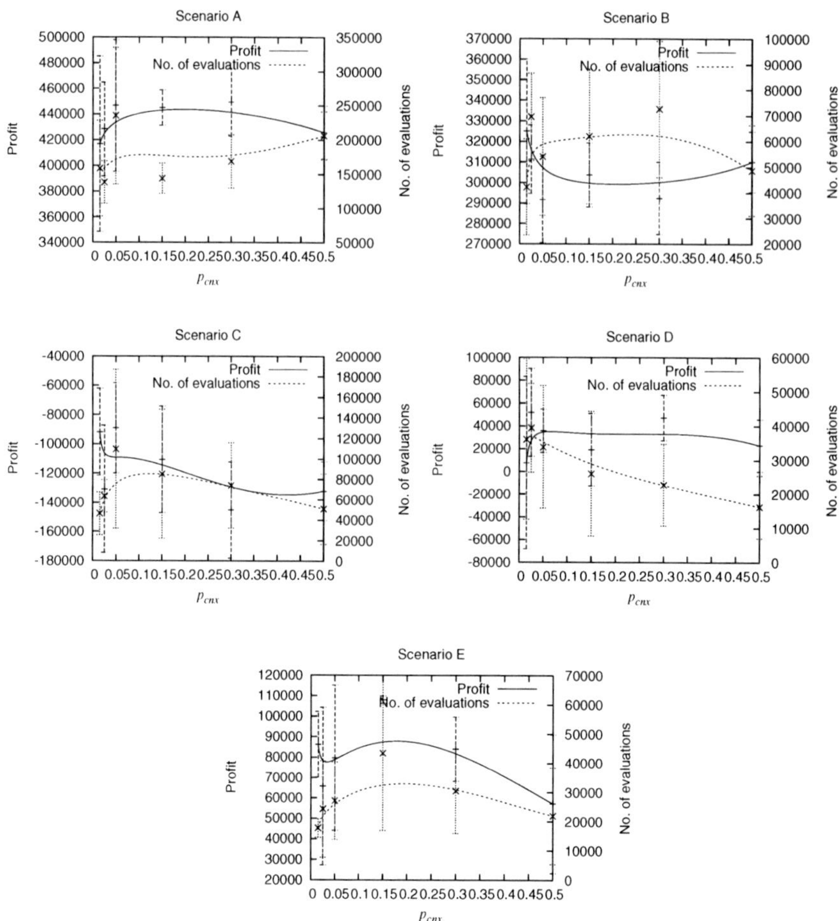

Fig. C.3 Calibration results for parameter p_{cnx}

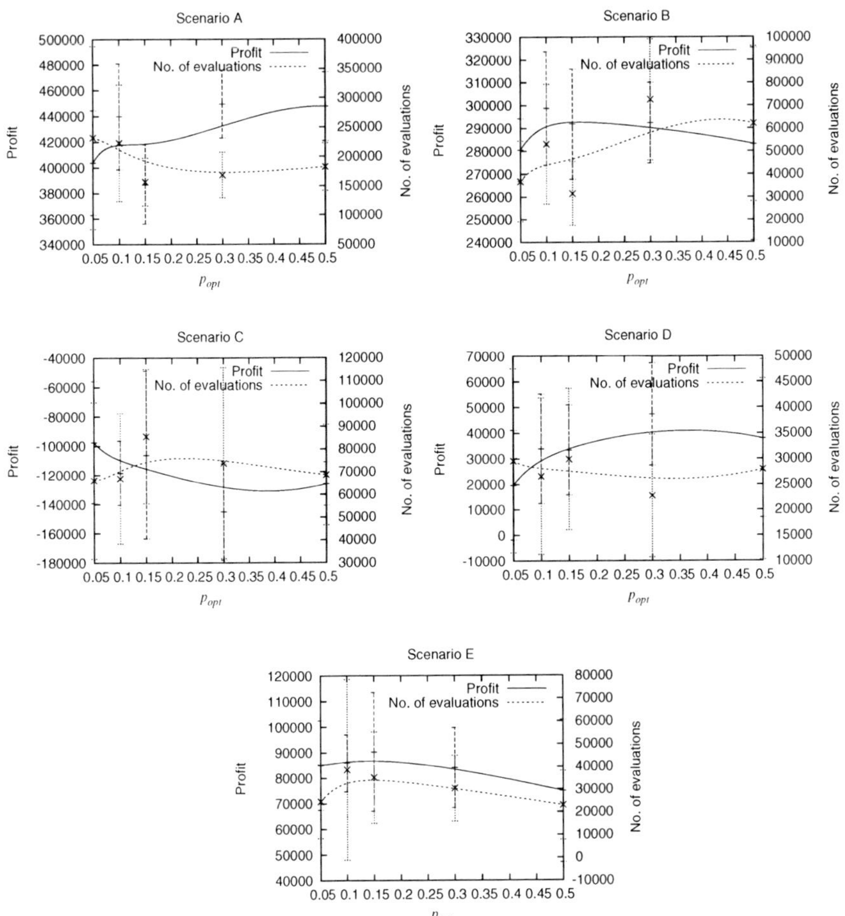

Fig. C.4 Calibration results for parameter p_{opt}

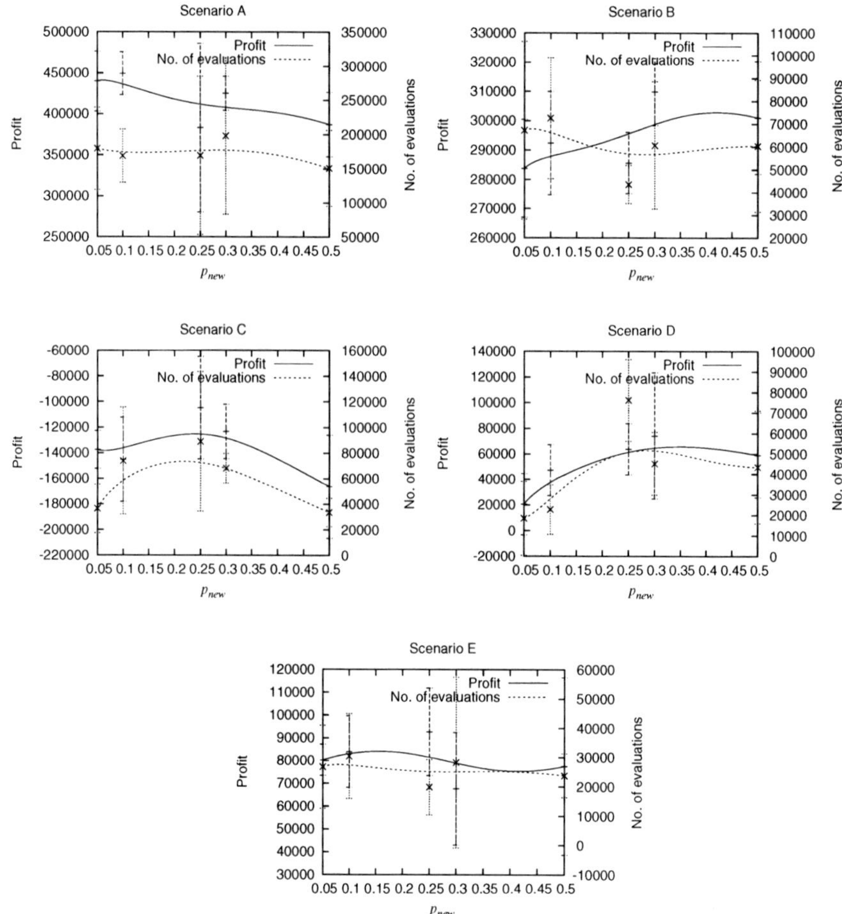

Fig. C.5 Calibration results for parameter p_{new}

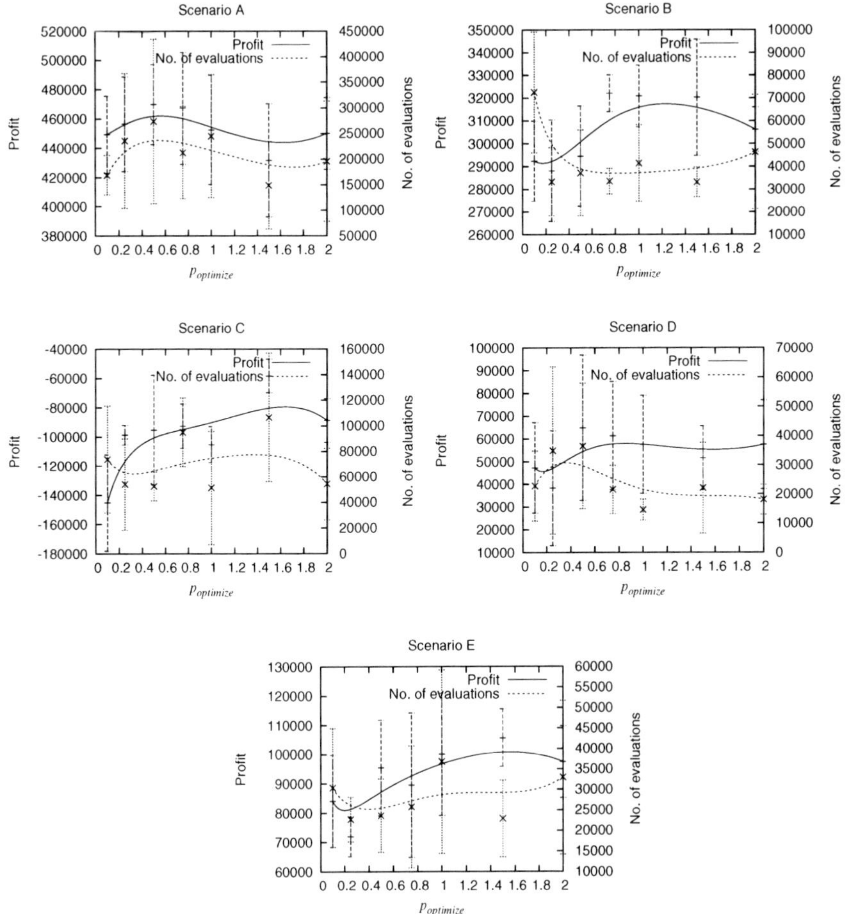

Fig. C.6 Calibration results for parameter $p_{optimize}$

C.1.2 Simultaneous Approach

C.1.2.1 Threshold Accepting

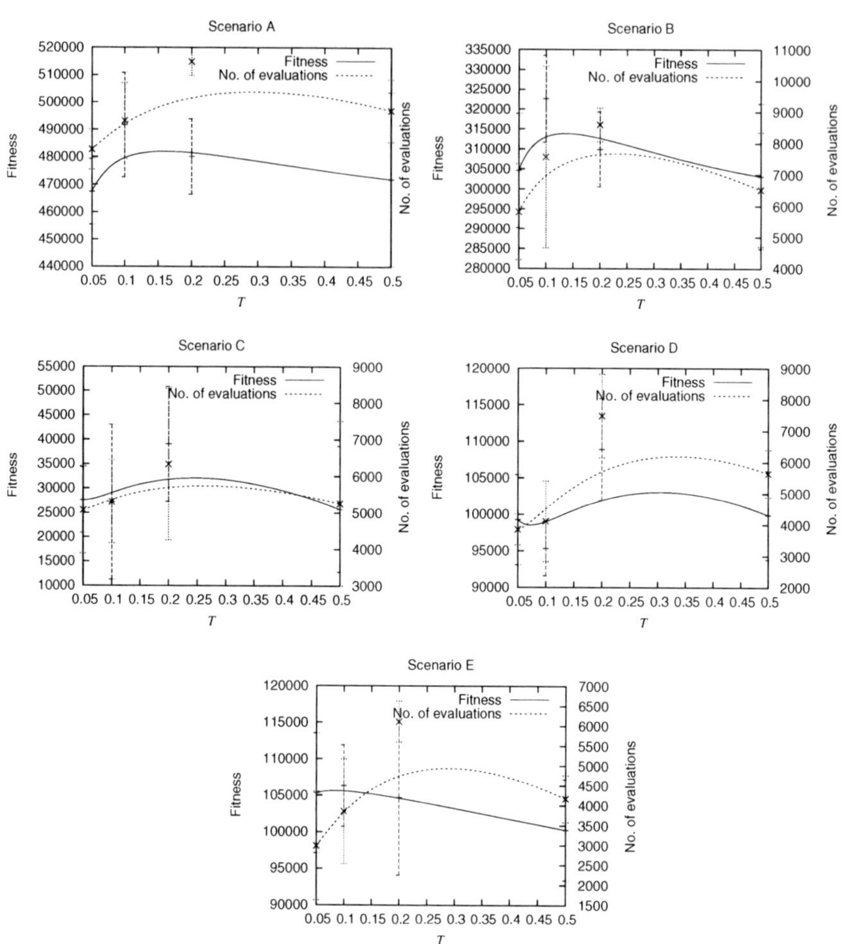

Fig. C.7 Calibration results for parameter T

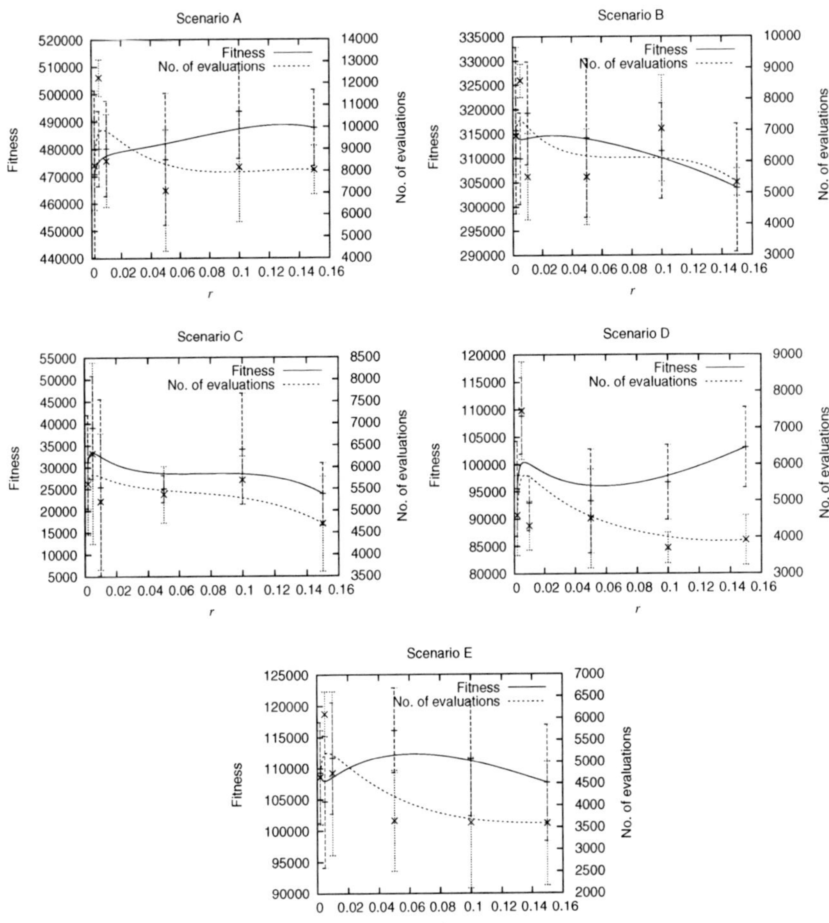

Fig. C.8 Calibration results for parameter r

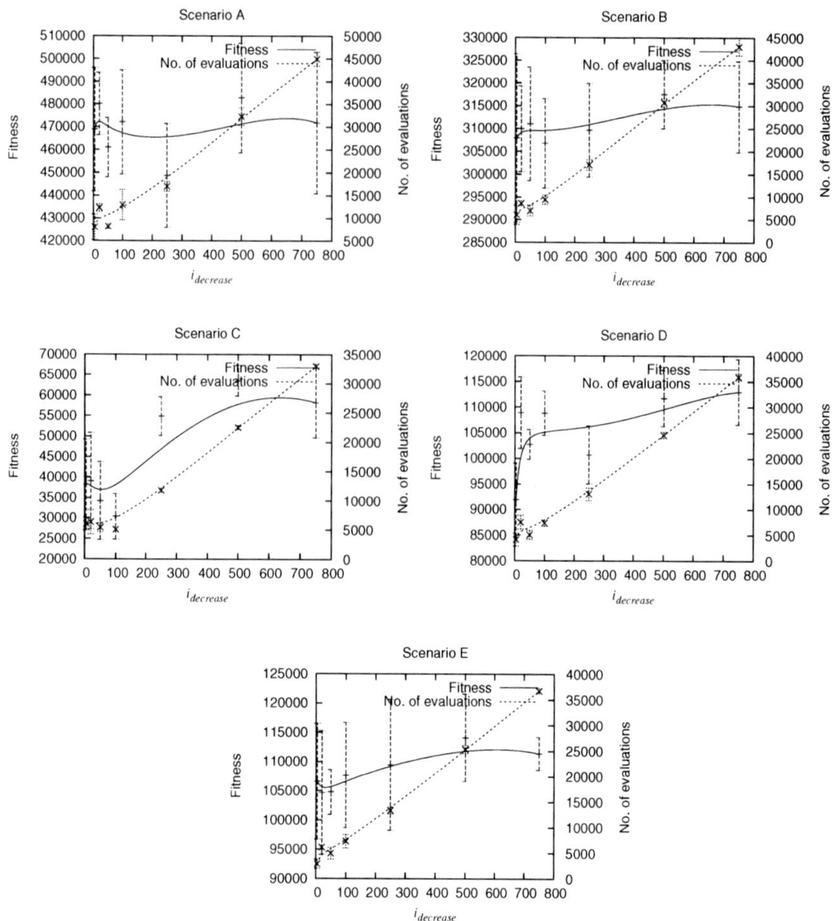

Fig. C.9 Calibration results for parameter $i_{decrease}$

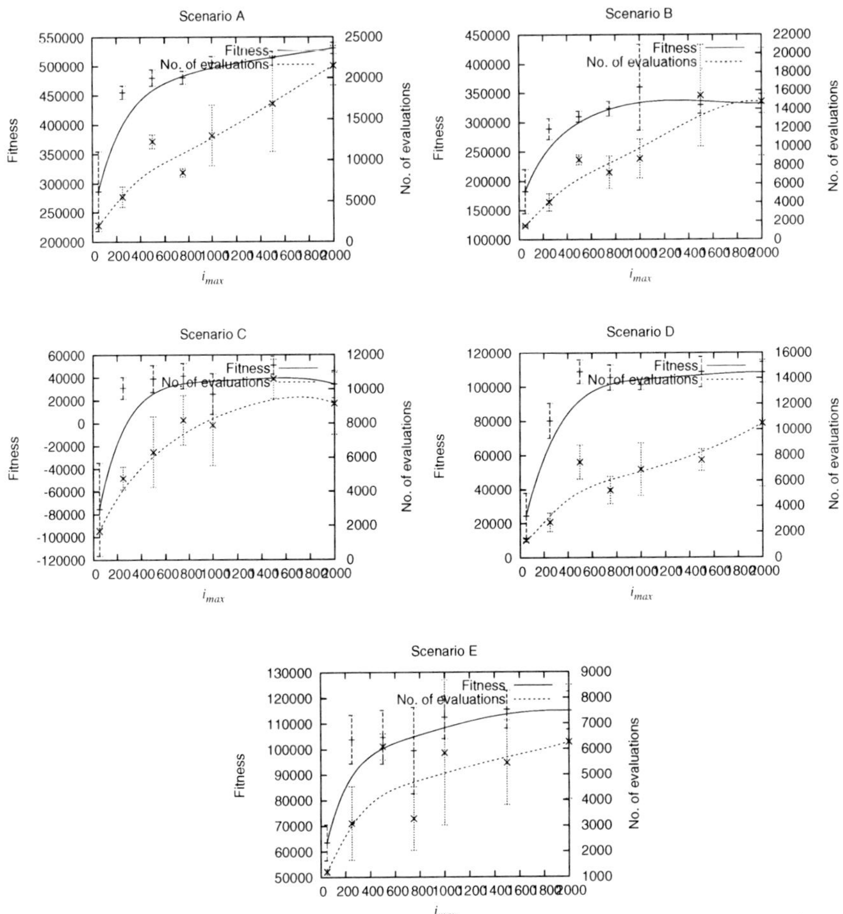

Fig. C.10 Calibration results for parameter i_{max}

C.1.2.2 Selecto-Recombinative Genetic Algorithm

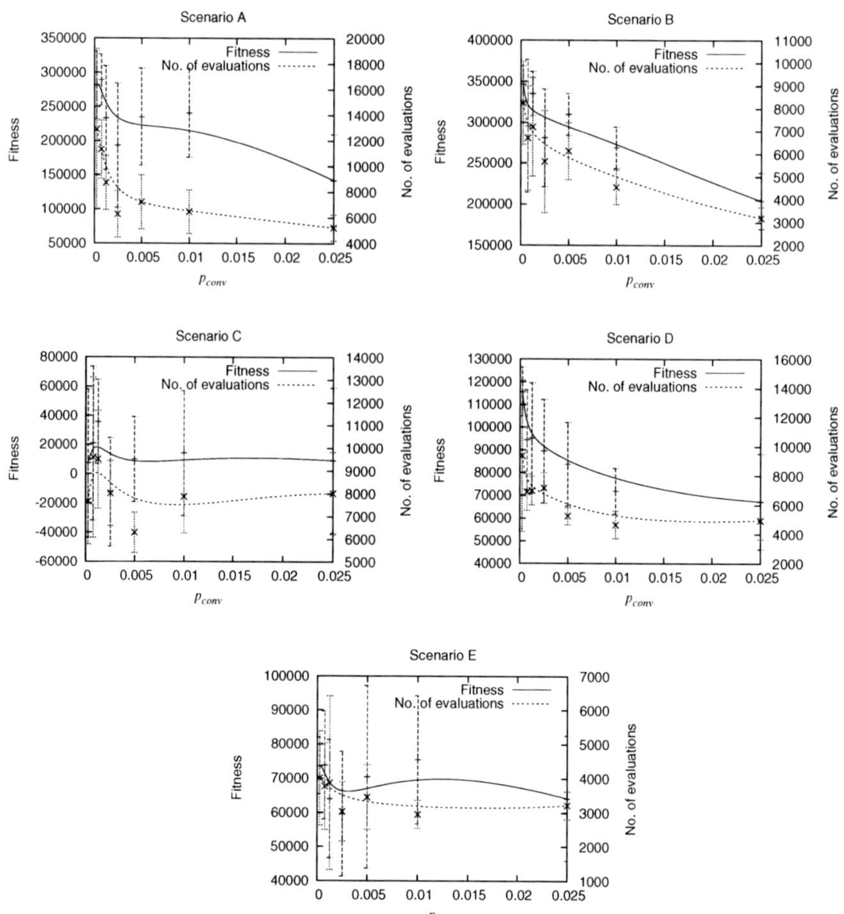

Fig. C.11 Calibration results for parameter p_{conv} of the selecto-recombinative genetic algorithm (rGA)

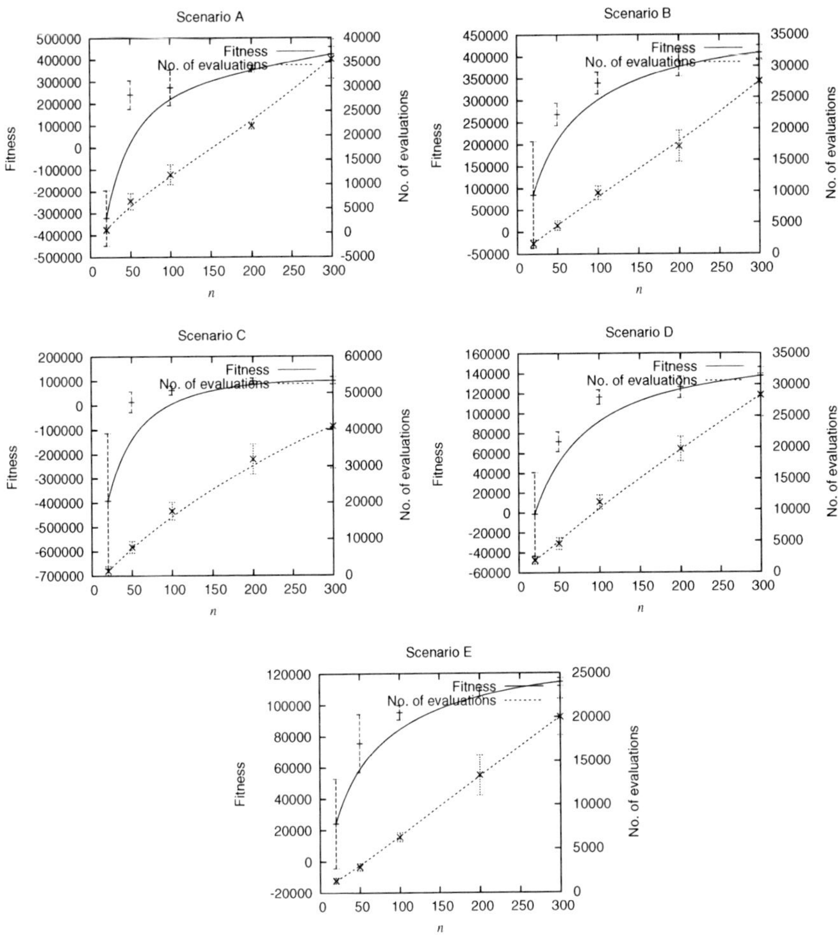

Fig. C.12 Calibration results for parameter n of the selecto-recombinative genetic algorithm (rGA)

C.1.2.3 Standard Genetic Algorithm

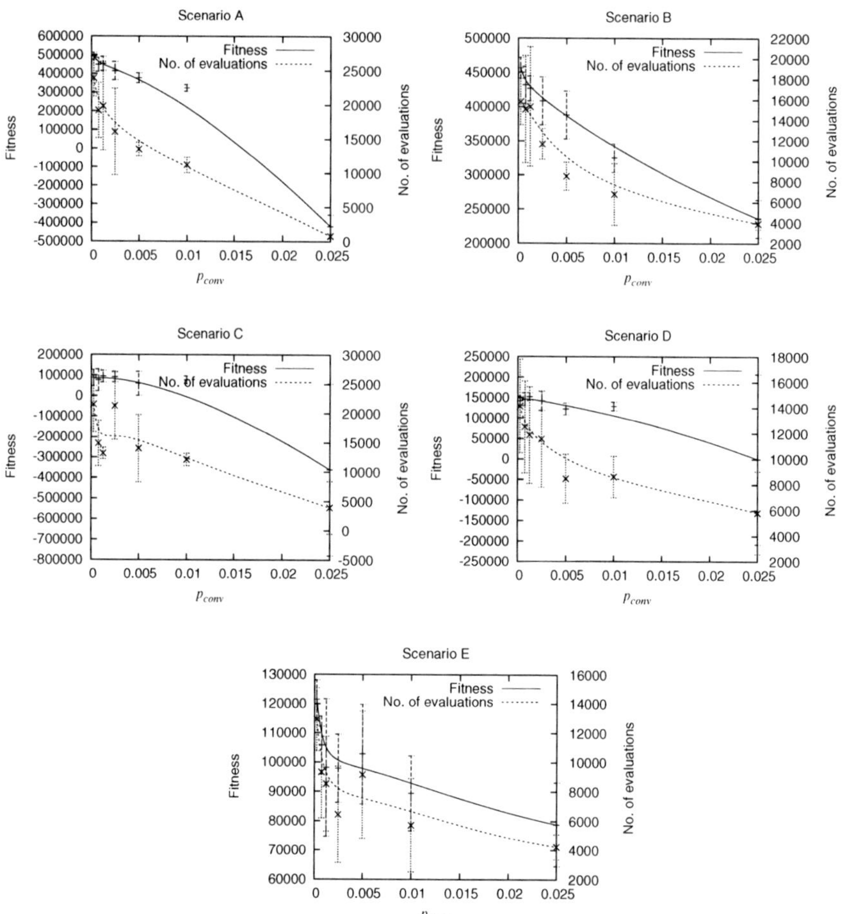

Fig. C.13 Calibration results for parameter p_{conv} of the standard genetic algorithm (GA)

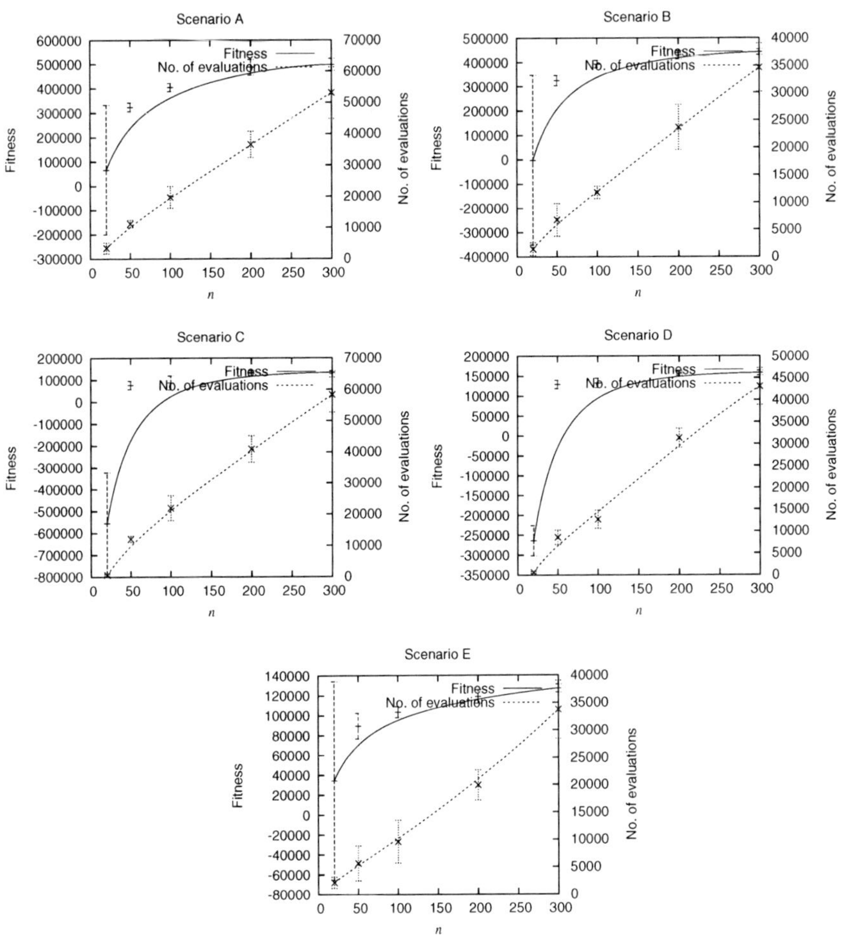

Fig. C.14 Calibration results for parameter n of the standard genetic algorithm (GA)

C.1.2.4 Strategy Selection

Table C.1 Resulting fitness using threshold accepting (TA)

	Scenario				
Run	A	B	C	D	E
1	489,465	328,156	63,590	103,849	111,929
2	516,468	320,000	60,927	94,525	120,079
3	483,236	332,933	69,712	108,726	117,354
4	521,768	323,468	63,246	110,793	125,325
5	527,743	319,994	64,205	108,187	120,771

Table C.2 Required number of fitness evaluations using threshold accepting (TA)

Run	A	B	C	D	E
			Scenario		
1	56,796	49,858	36,939	40,421	40,894
2	54,026	53,172	37,695	44,806	41,647
3	54,001	49,251	40,854	40,345	41,522
4	57,903	47,200	37,505	41,980	42,688
5	61,352	48,089	37,218	39,196	40,046

Table C.3 Resulting fitness using the selecto-recombinative genetic algorithm (rGA)

Run	A	B	C	D	E
			Scenario		
1	347,812	398,837	114,085	139,631	119,946
2	435,642	421,190	107,070	147,816	104,347
3	459,327	402,376	100,477	134,438	101,620
4	421,799	413,799	90,472	144,163	109,115
5	384,980	429,133	106,613	148,210	114,292

Table C.4 Required number of fitness evaluations using the selecto-recombinative genetic algorithm (rGA)

Run	A	B	C	D	E
			Scenario		
1	21,052	23,481	28,793	22,206	19,318
2	37,647	25,137	29,444	35,411	13,957
3	33,913	20,309	26,949	18,746	16,287
4	34,622	23,886	36,193	33,509	15,421
5	23,044	28,970	30,216	34,656	16,715

Table C.5 Resulting fitness using the genetic algorithm (GA)

Run	A	B	C	D	E
			Scenario		
1	558,384	488,774	139,998	171,750	113,665
2	549,128	482,493	127,728	152,367	125,594
3	600,414	486,785	110,686	163,462	132,223
4	584,197	488,329	138,894	161,929	126,265
5	566,935	482,492	134,616	178,271	131,651

Table C.6 Required number of fitness evaluations using the genetic algorithm (GA)

			Scenario		
Run	A	B	C	D	E
1	63,552	41,721	93,621	48,037	11,861
2	57,362	50,254	55,843	32,903	37,659
3	75,436	48,409	61,923	39,372	38,005
4	79,831	46,607	70,314	39,677	56,353
5	72,979	47,555	66,051	57,858	39,024

C.2 Analysis

C.2.1 Sequential Approach

C.2.1.1 Solution Quality

Table C.7 Profit of airline schedules constructed with the sequential planning approach

			Scenario		
Run	A	B	C	D	E
1	348,903	330,092	-33,872	41,249	87,012
2	483,453	311,560	-52,610	34,716	103,510
3	457,167	328,166	-75,166	64,826	99,463
4	502,159	350,320	-54,032	82,175	94,027
5	461,464	309,495	-85,149	36,280	104,857
Average	450,629	325,927	-60,166	51,849	97,774
Stand. dev.	59,668	16,536	20,219	20,835	7,345

Table C.8 Seat load factors (SLF) of airline schedules constructed with the sequential planning approach

			Scenario		
Run	A	B	C	D	E
1	0.282	0.373	0.154	0.382	0.293
2	0.310	0.378	0.175	0.386	0.246
3	0.307	0.332	0.196	0.394	0.231
4	0.306	0.388	0.176	0.353	0.282
5	0.294	0.374	0.158	0.419	0.285
Average	0.300	0.369	0.172	0.387	0.267
Stand. dev.	0.012	0.021	0.017	0.023	0.027

Table C.9 Numbers of passengers of airline schedules constructed with the sequential planning approach

Run	Scenario				
	A	B	C	D	E
1	5,208	3,542	1,720	2,158	2,270
2	6,276	3,416	1,647	2,144	1,832
3	5,384	3,255	1,718	2,096	1,717
4	6,240	3,756	1,740	2,217	2,134
5	6,627	3,495	1,584	2,106	2,332
Average	5,947	3,493	1,682	2,144	2,057
Stand. dev.	616	183	65	48	271

Table C.10 Numbers of flights of airline schedules constructed with the sequential planning approach

Run	Scenario				
	A	B	C	D	E
1	126	101	56	72	52
2	138	92	61	75	44
3	126	84	61	70	44
4	140	105	61	78	50
5	151	101	65	68	55
Average	136.20	96.60	60.80	72.60	49.00
Stand. dev.	10.55	8.50	3.19	3.97	4.90

Table C.11 Numbers of fitness evaluations required by the sequential planning approach

Run	Scenario				
	A	B	C	D	E
1	118,490	27,455	247,118	46,919	91,260
2	493,759	28,496	63,574	37,720	25,815
3	70,099	38,160	128,822	20,719	12,644
4	79,425	55,600	242,930	16,790	80,939
5	302,656	34,150	90,645	39,175	32,448
Average	212,886	36,772	154,618	32,265	48,621
Stand. dev.	183,070	11,386	85,735	12,894	35,137

Table C.12 Total numbers of fitness evaluations required by the sequential planning approach

	Scenario				
Run	A	B	C	D	E
1	255,291	104,883	414,296	50,558	98,806
2	499,332	117,453	137,323	41,298	109,869
3	307,674	124,912	132,350	40,023	75,588
4	84,546	140,852	399,100	55,591	96,906
5	698,906	72,351	219,072	64,203	35,676
Average	369,150	112,090	260,428	50,335	83,369
Stand. dev.	236,341	25,747	137,999	10,100	29,406

Table C.13 Numbers of iterations required by the sequential planning approach

	Scenario				
Run	A	B	C	D	E
1	13	25	52	11	26
2	35	24	19	5	23
3	16	24	20	7	20
4	5	30	55	12	20
5	26	15	34	12	12
Average	19	24	36	9	20
Stand. dev.	12	5	17	3	5

C.2.1.2   Solution Process

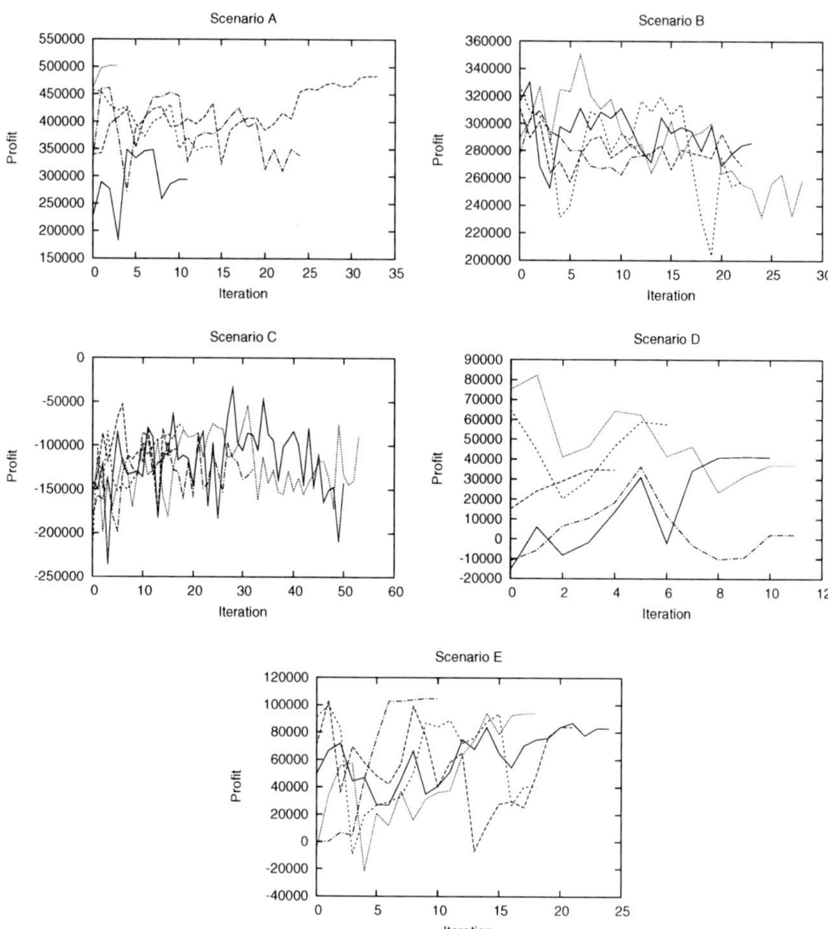

Fig. C.15 Trend of profits

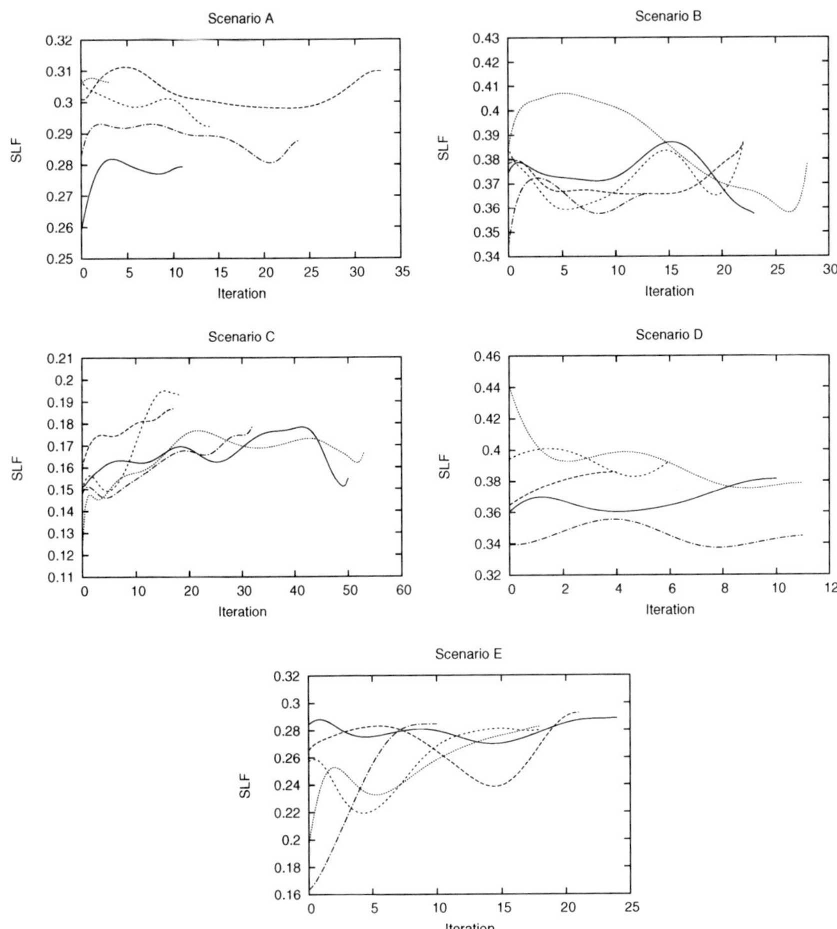

Fig. C.16 Trend of seat load factors (SLF)

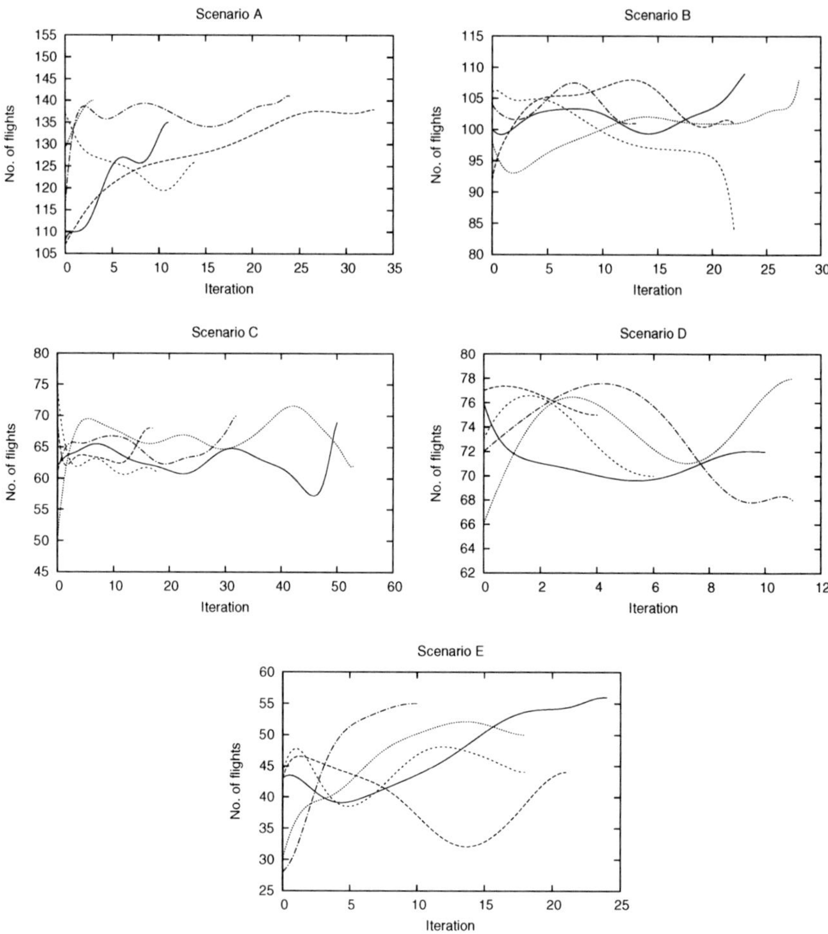

Fig. C.17 Trend of numbers of flights

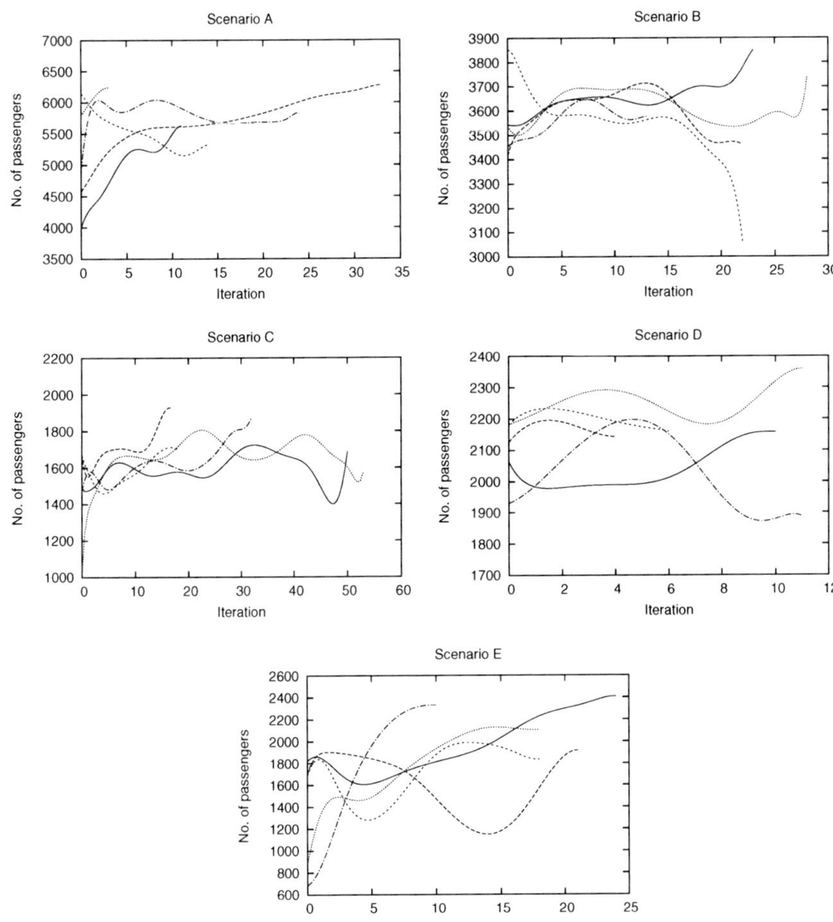

Fig. C.18 Trend of numbers of passengers

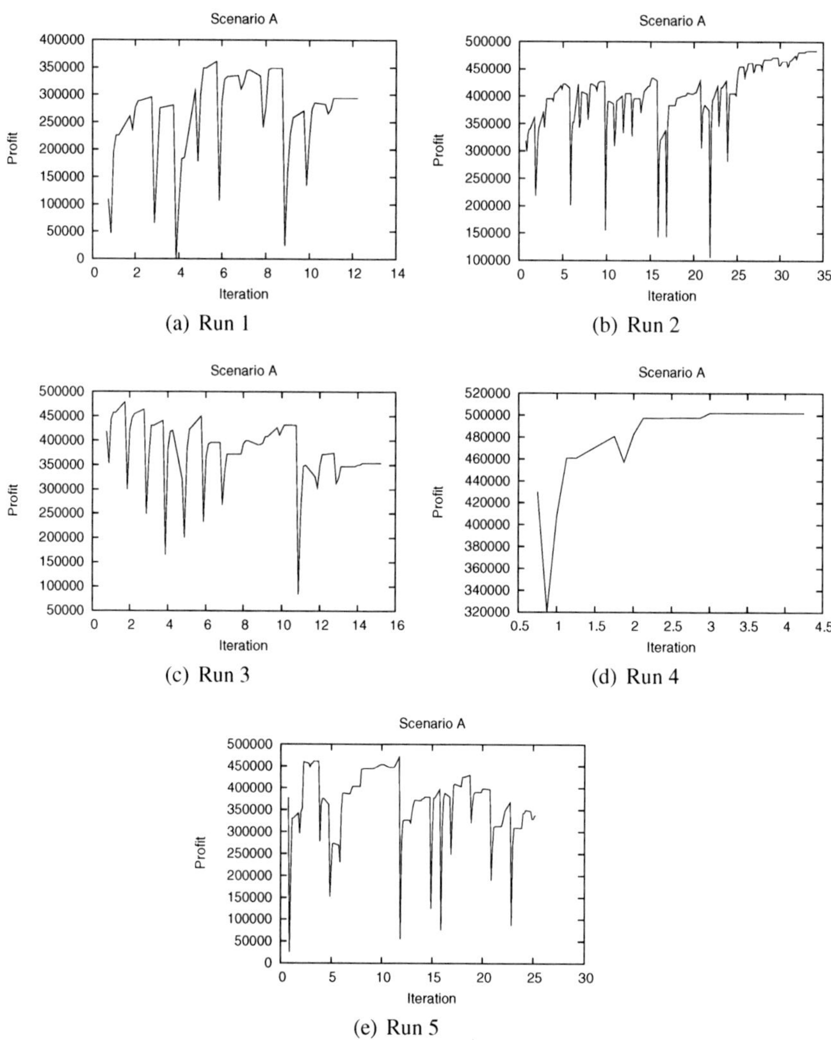

(a) Run 1

(b) Run 2

(c) Run 3

(d) Run 4

(e) Run 5

Fig. C.19 Profit contribution by individual solution steps (scenario A)

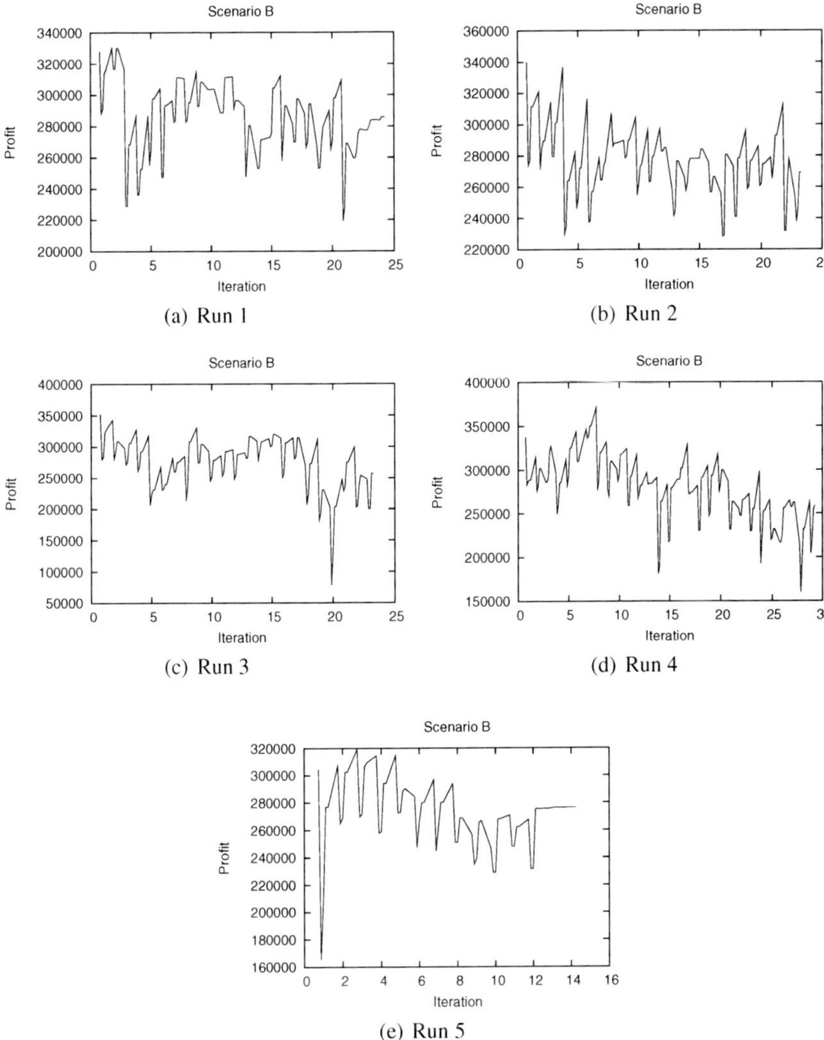

Fig. C.20 Profit contribution by individual solution steps (scenario B)

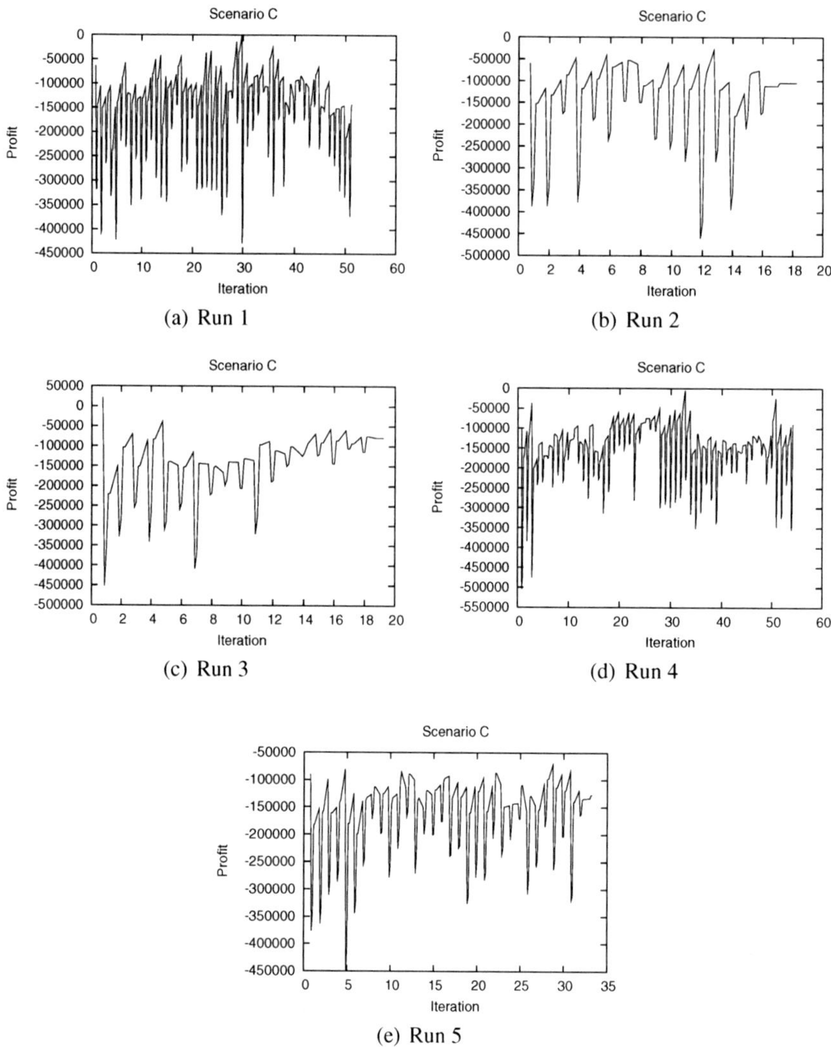

(a) Run 1

(b) Run 2

(c) Run 3

(d) Run 4

(e) Run 5

Fig. C.21 Profit contribution by individual solution steps (scenario C)

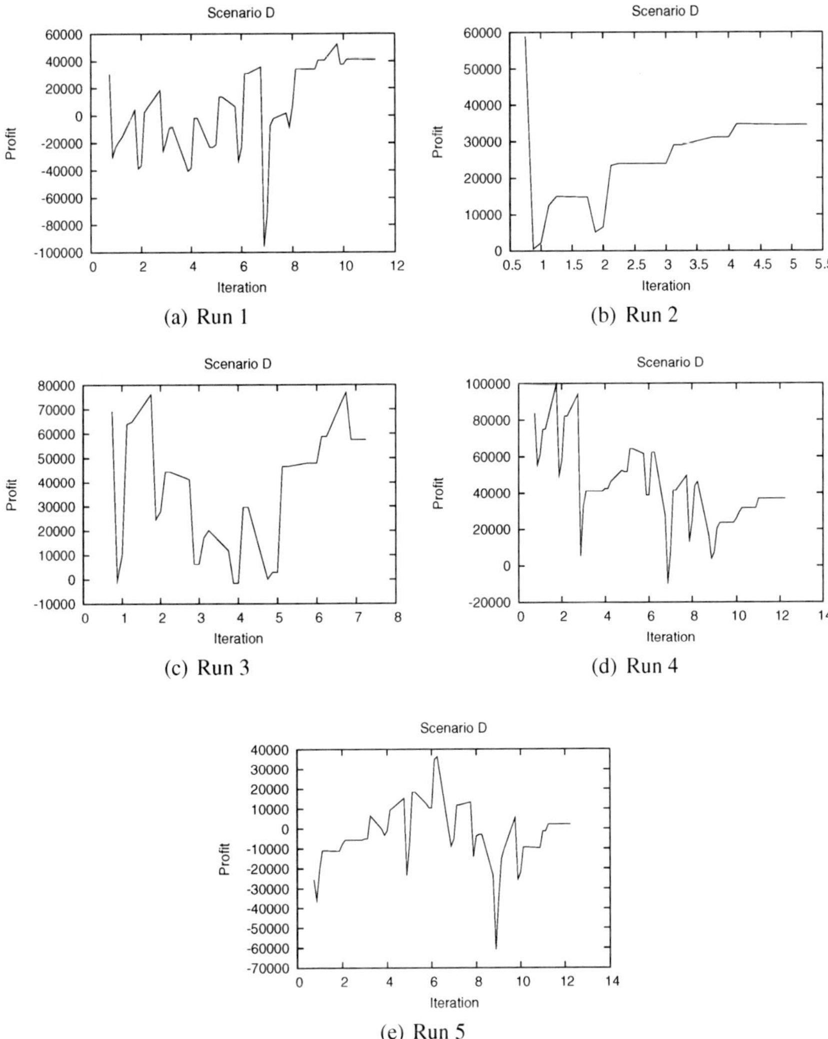

Fig. C.22 Profit contribution by individual solution steps (scenario D)

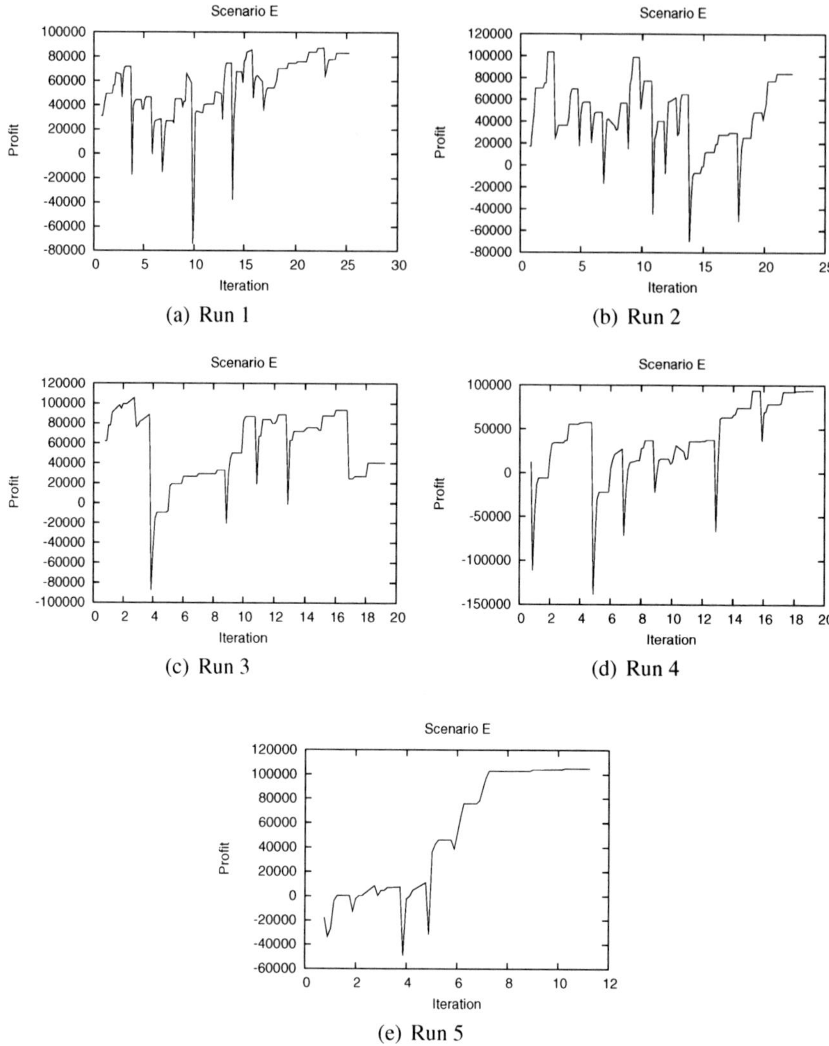

Fig. C.23 Profit contribution by individual solution steps (scenario E)

C.2.2 *Simultaneous Approach*

C.2.2.1 Solution Quality

Table C.14 Fitness of airline schedules constructed with the simultaneous planning approach

			Scenario		
Run	A	B	C	D	E
1	558,384	488,774	139,998	171,750	113,665
2	549,128	482,493	127,728	152,367	125,594
3	600,414	486,785	110,686	163,462	132,223
4	584,197	488,329	138,894	161,929	126,265
5	566,935	482,492	134,616	178,271	131,651
Average	571,812	485,775	130,384	165,556	125,880
Stand. dev.	20,556	3,086	12,019	9,894	7,466

Table C.15 Seat load factors (SLF) of airline schedules constructed with the simultaneous planning approach

			Scenario		
Run	A	B	C	D	E
1	0.336	0.500	0.216	0.497	0.306
2	0.335	0.512	0.210	0.478	0.303
3	0.376	0.490	0.231	0.482	0.327
4	0.326	0.502	0.222	0.494	0.316
5	0.346	0.498	0.217	0.523	0.327
Average	0.344	0.501	0.219	0.495	0.316
Stand. dev.	0.019	0.008	0.008	0.018	0.011

Table C.16 Numbers of passengers of airline schedules constructed with the simultaneous planning approach

			Scenario		
Run	A	B	C	D	E
1	5,465	4,551	2,529	2,575	1,459
2	5,480	4,304	2,665	2,410	1,759
3	5,531	4,448	2,485	2,532	2,047
4	5,732	4,449	2,524	2,492	1,787
5	5,467	4,457	2,553	2,454	1,951
Average	5,535	4,442	2,551	2,493	1,801
Stand. dev.	113	88	68	65	225

Table C.17 Numbers of flights of airline schedules constructed with the simultaneous planning approach

			Scenario		
Run	A	B	C	D	E
1	113	130	68	74	32
2	116	120	74	72	39
3	112	126	68	75	42
4	117	124	68	72	38
5	110	127	70	67	40
Average	113.60	125.40	69.60	72.00	38.20
Stand. dev.	2.88	3.71	2.61	3.08	3.77

Table C.18 Numbers of fitness evaluations required by the simultaneous planning approach

			Scenario		
Run	A	B	C	D	E
1	63,552	41,721	93,621	48,037	11,861
2	57,362	50,254	55,843	32,903	37,659
3	75,436	48,409	61,923	39,372	38,005
4	79,831	46,607	70,314	39,677	56,353
5	72,979	47,555	66,051	57,858	39,024
Average	69,832	46,909	69,550	43,569	36,580
Stand. dev.	9,168	3,196	14,477	9,626	15,899

Table C.19 Total numbers of fitness evaluations required by the simultaneous planning approach

			Scenario		
Run	A	B	C	D	E
1	64,511	43,472	94,379	48,662	12,331
2	57,787	50,284	57,042	35,310	39,996
3	77,329	49,441	63,905	40,892	40,545
4	80,636	47,312	73,650	40,132	57,400
5	73,352	48,778	67,271	58,610	43,140
Average	70,723	47,857	71,249	44,721	38,682
Stand. dev.	9,417	2,682	14,251	9,120	16,354

C.2.2.2 Solution Process

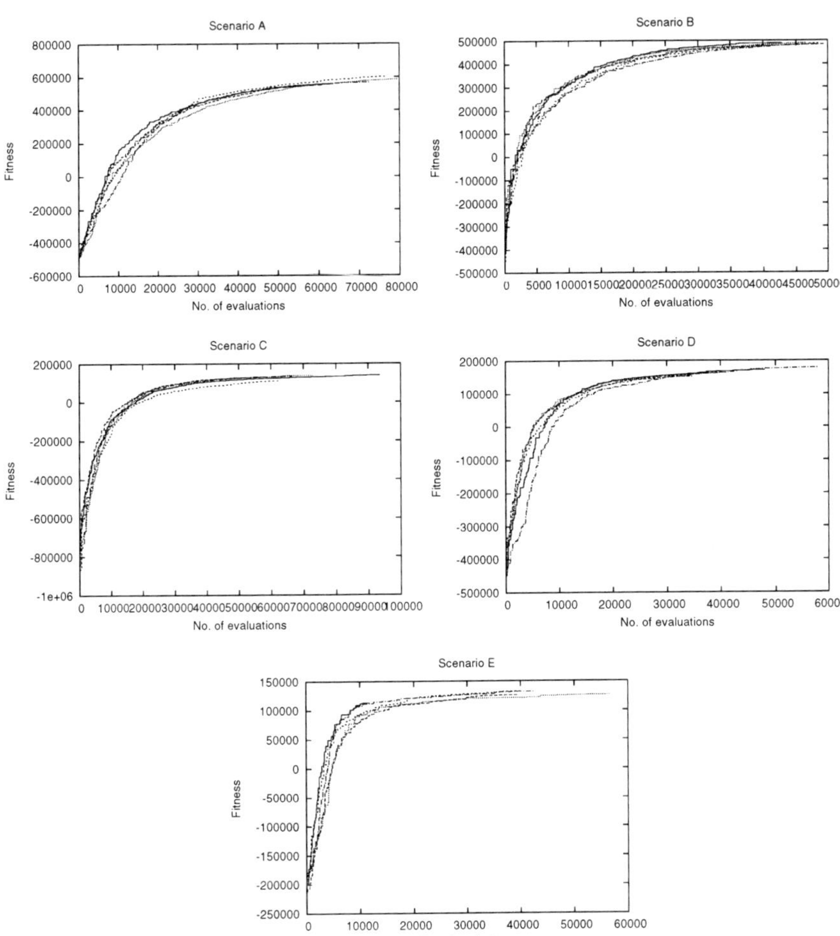

Fig. C.24 Trend of fitness values

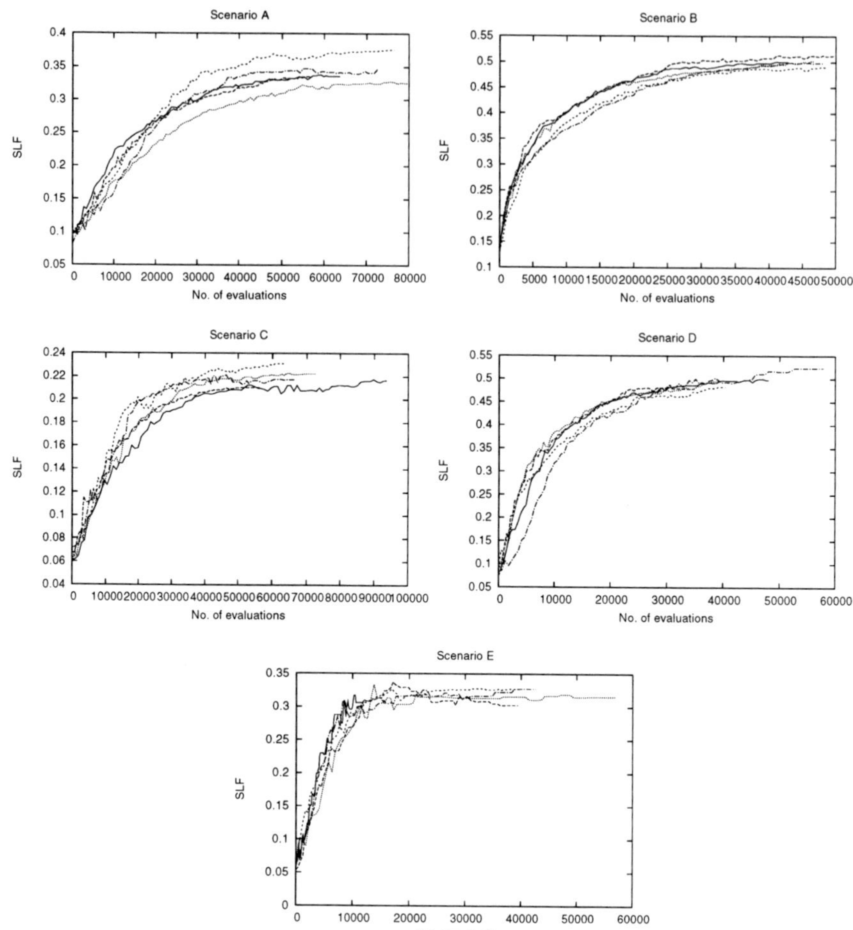

Fig. C.25 Trend of seat load factors (SLF)

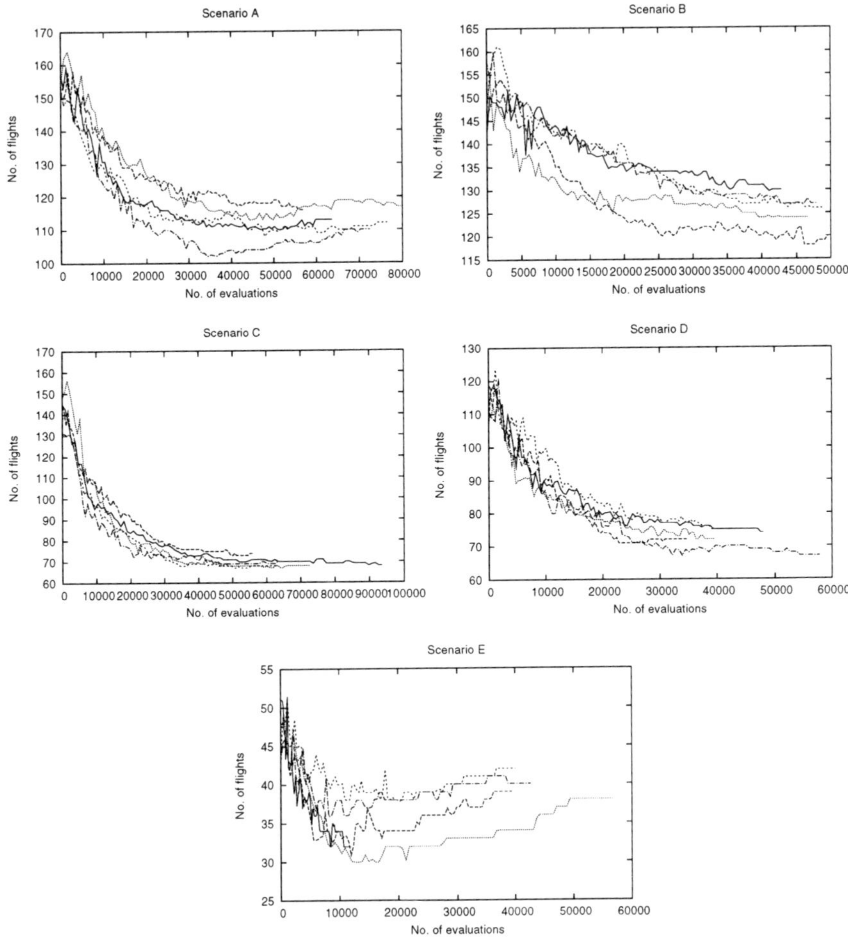

Fig. C.26 Trend of numbers of flights

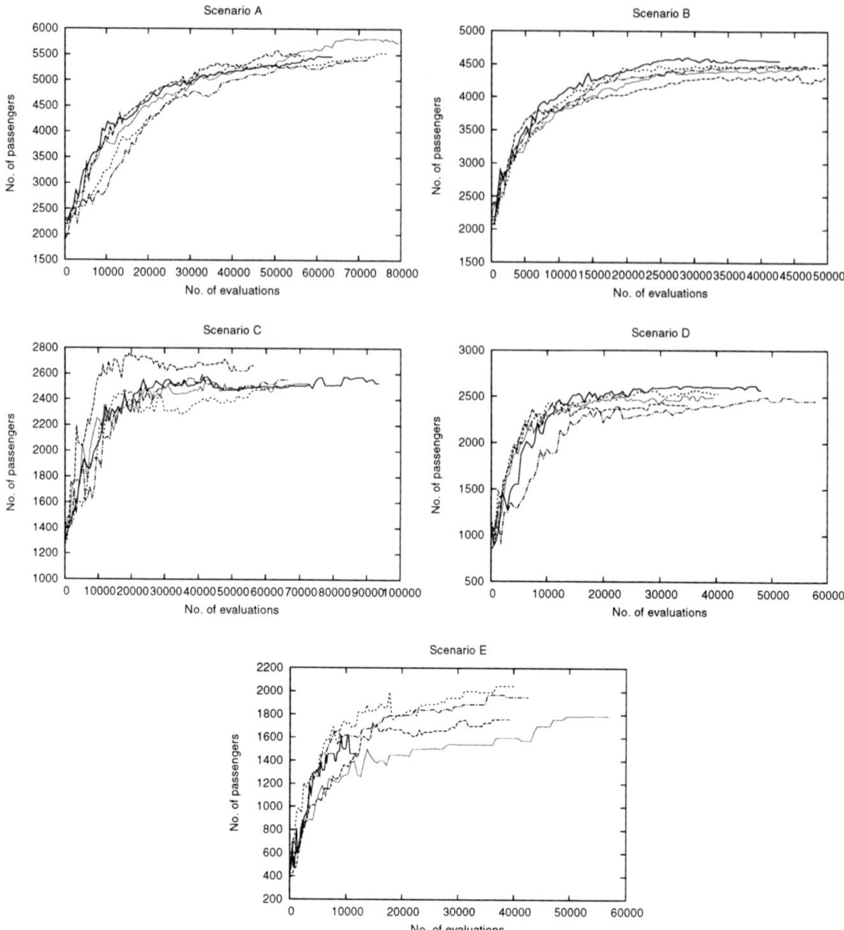

Fig. C.27 Trend of numbers of passengers

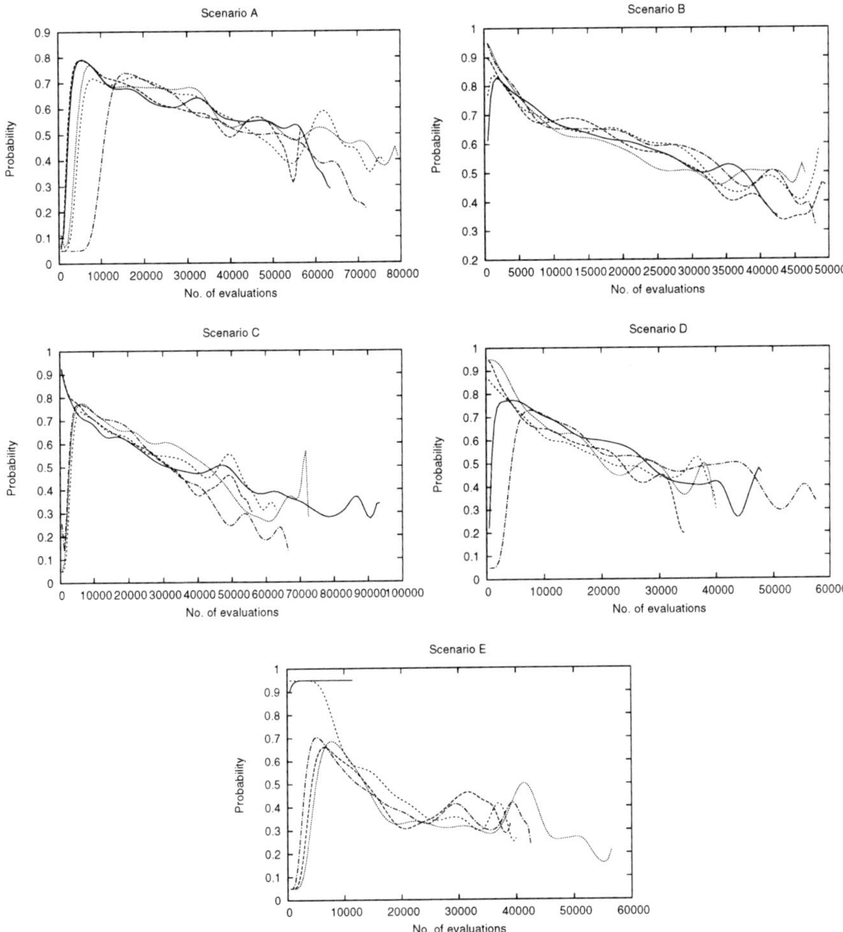

Fig. C.28 Application probability of recombination-based operators

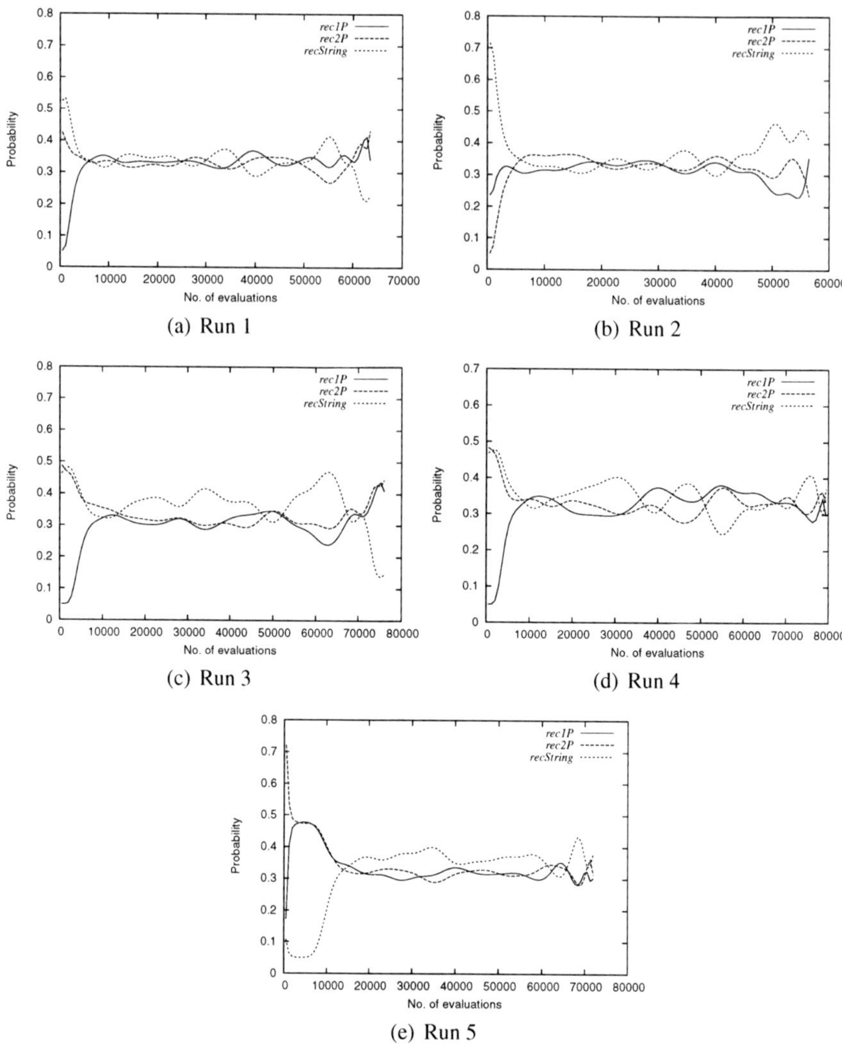

(a) Run 1

(b) Run 2

(c) Run 3

(d) Run 4

(e) Run 5

Fig. C.29 Application probability of the different variants of the recombination-based search operators (scenario A)

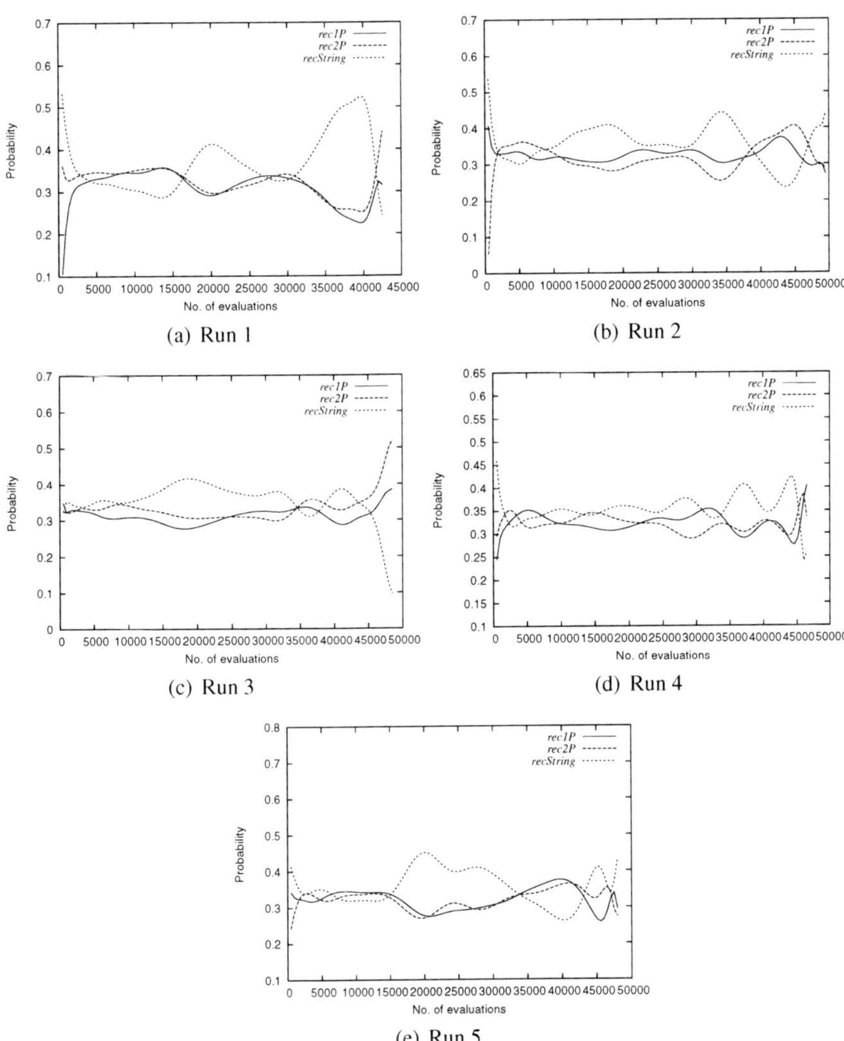

(a) Run 1

(b) Run 2

(c) Run 3

(d) Run 4

(e) Run 5

Fig. C.30 Application probability of the different variants of the recombination-based search operators (scenario B)

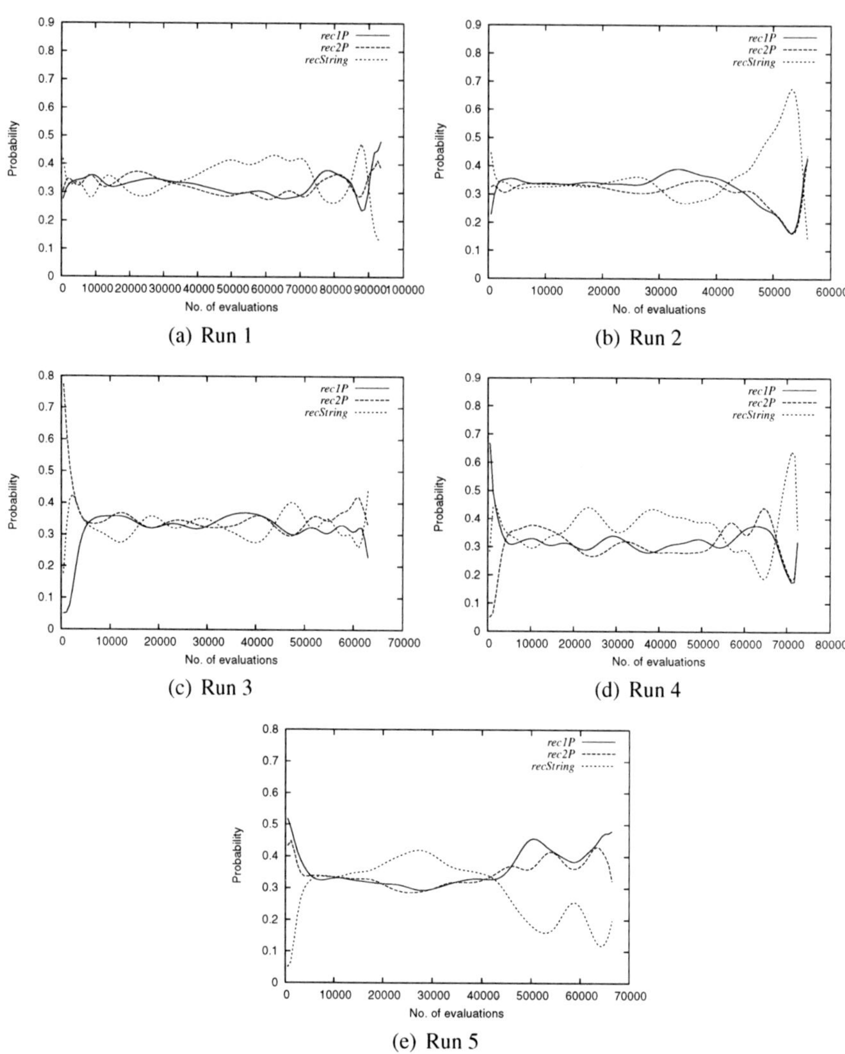

Fig. C.31 Application probability of the different variants of the recombination-based search operators (scenario C)

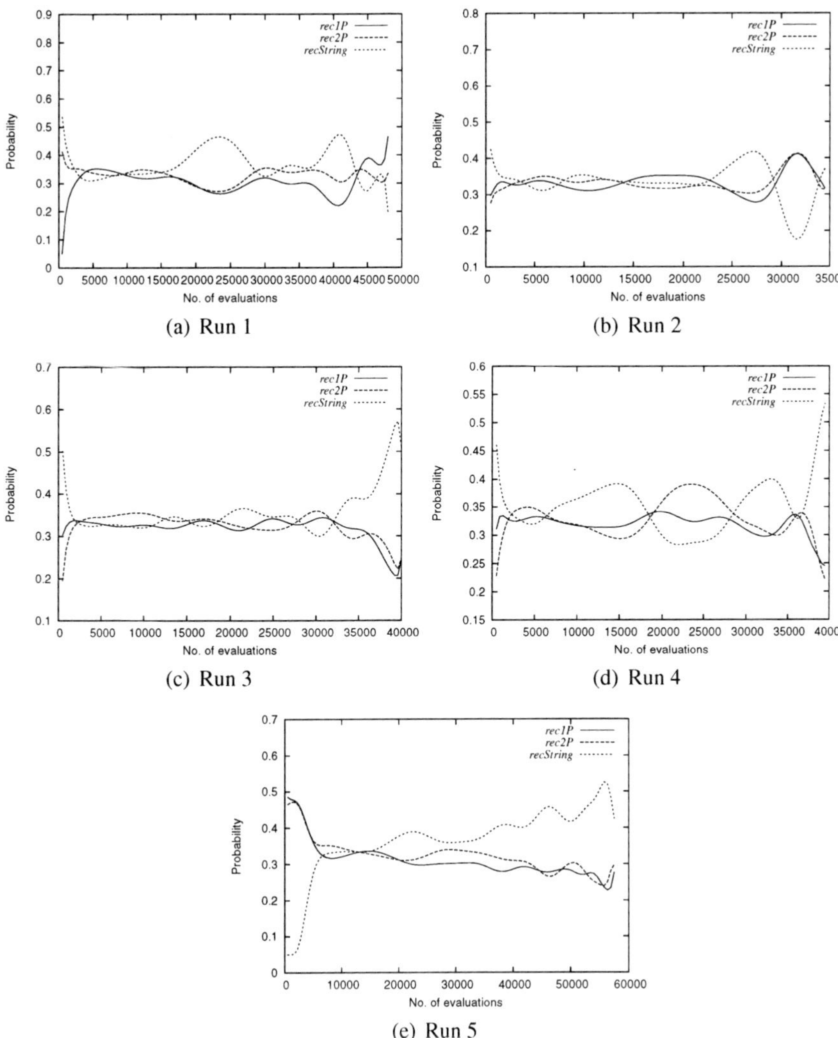

(a) Run 1

(b) Run 2

(c) Run 3

(d) Run 4

(e) Run 5

Fig. C.32 Application probability of the different variants of the recombination-based search operators (scenario D)

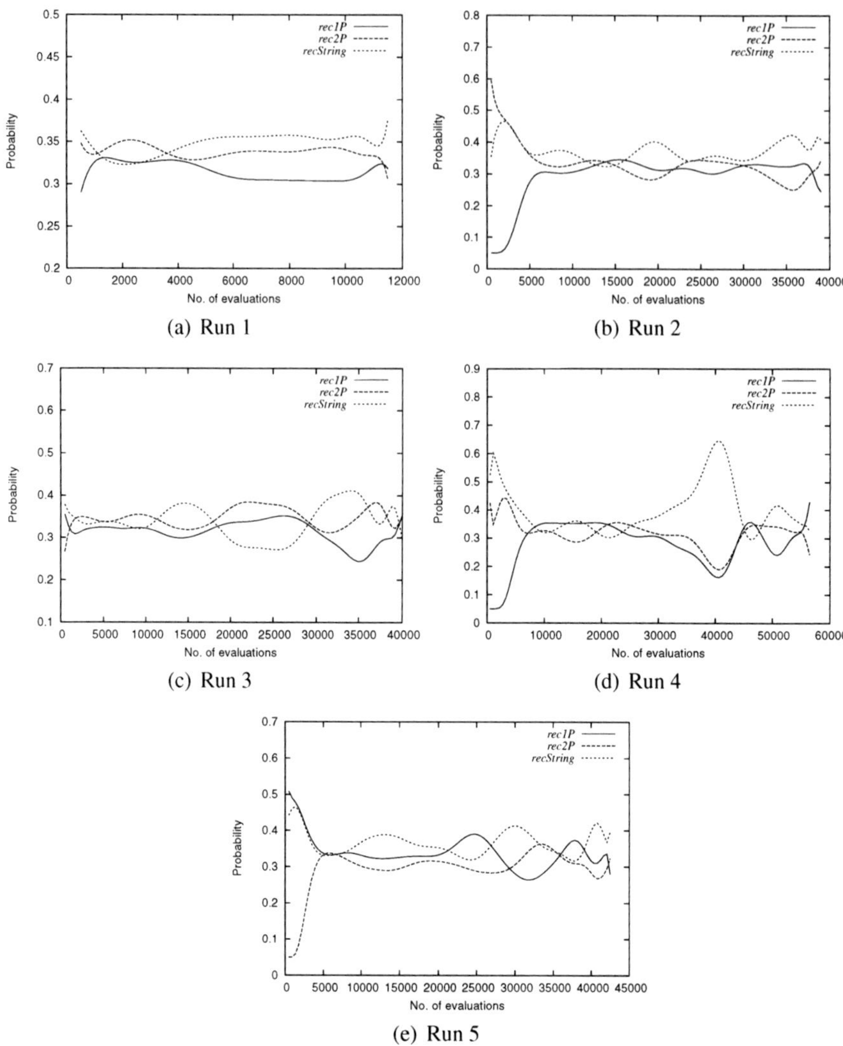

Fig. C.33 Application probability of the different variants of the recombination-based search operators (scenario E)

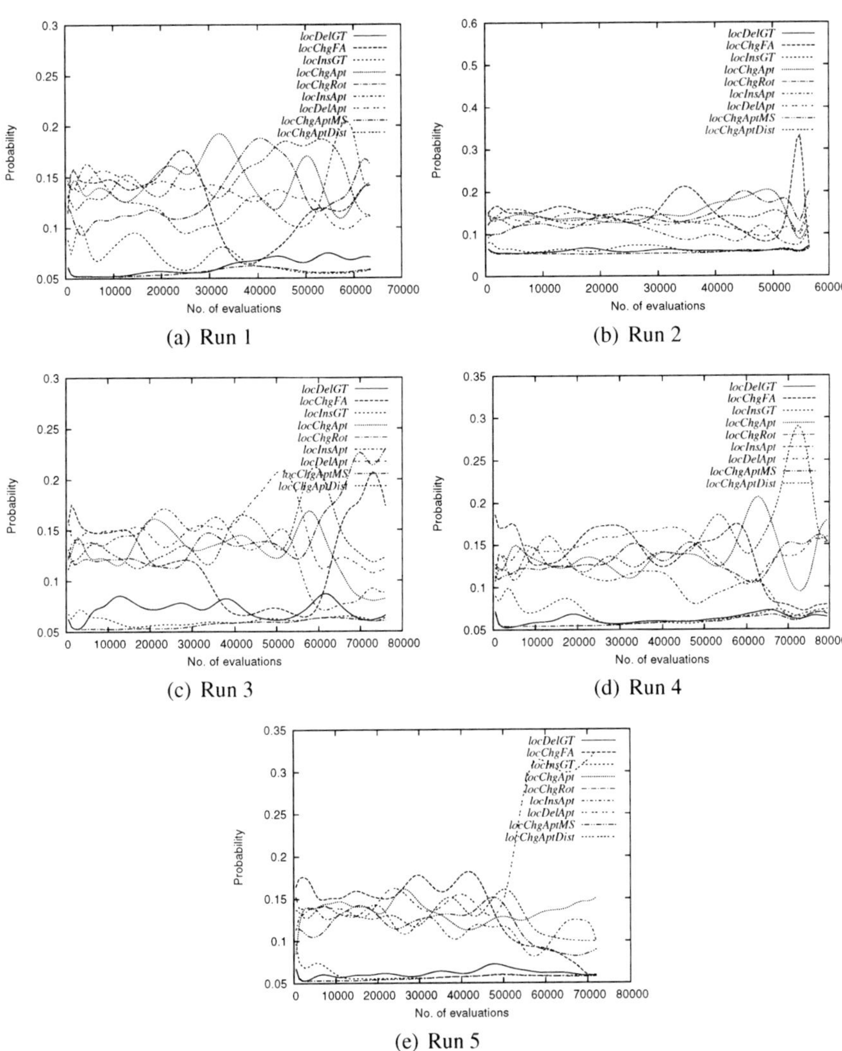

(a) Run 1

(b) Run 2

(c) Run 3

(d) Run 4

(e) Run 5

Fig. C.34 Application probability of the different variants of the local search operators (scenario A)

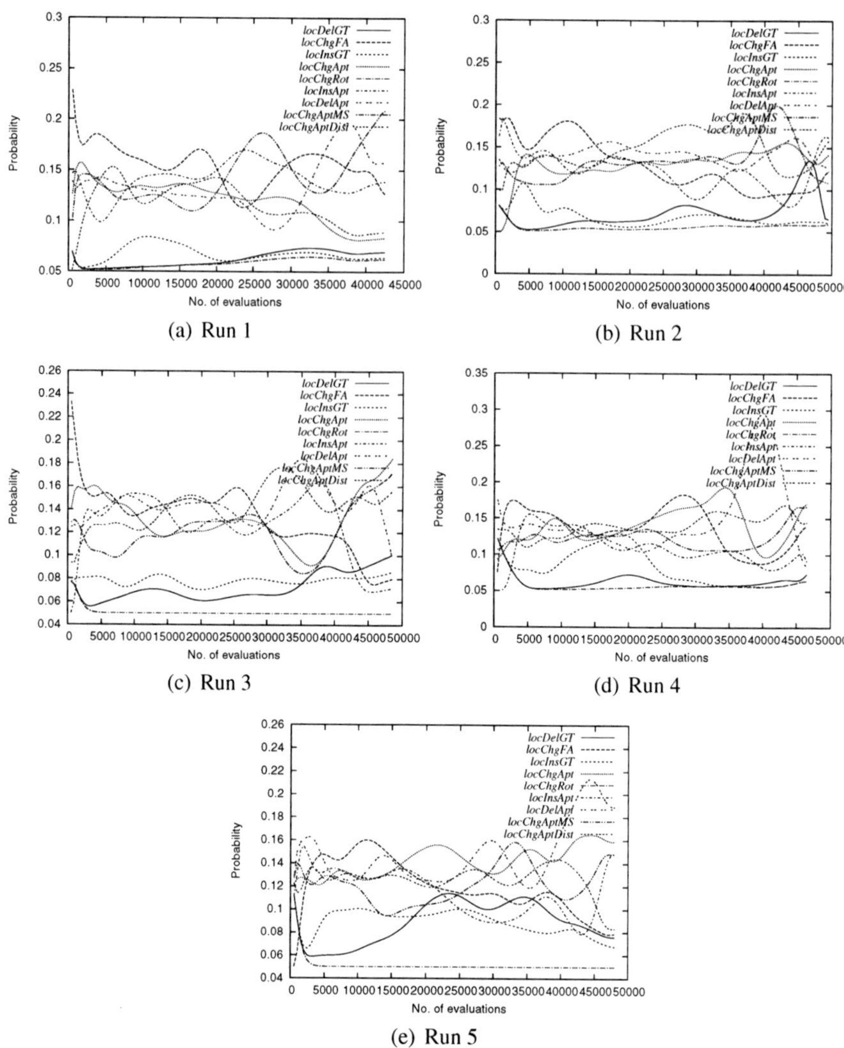

(a) Run 1

(b) Run 2

(c) Run 3

(d) Run 4

(e) Run 5

Fig. C.35 Application probability of the different variants of the local search operators (scenario B)

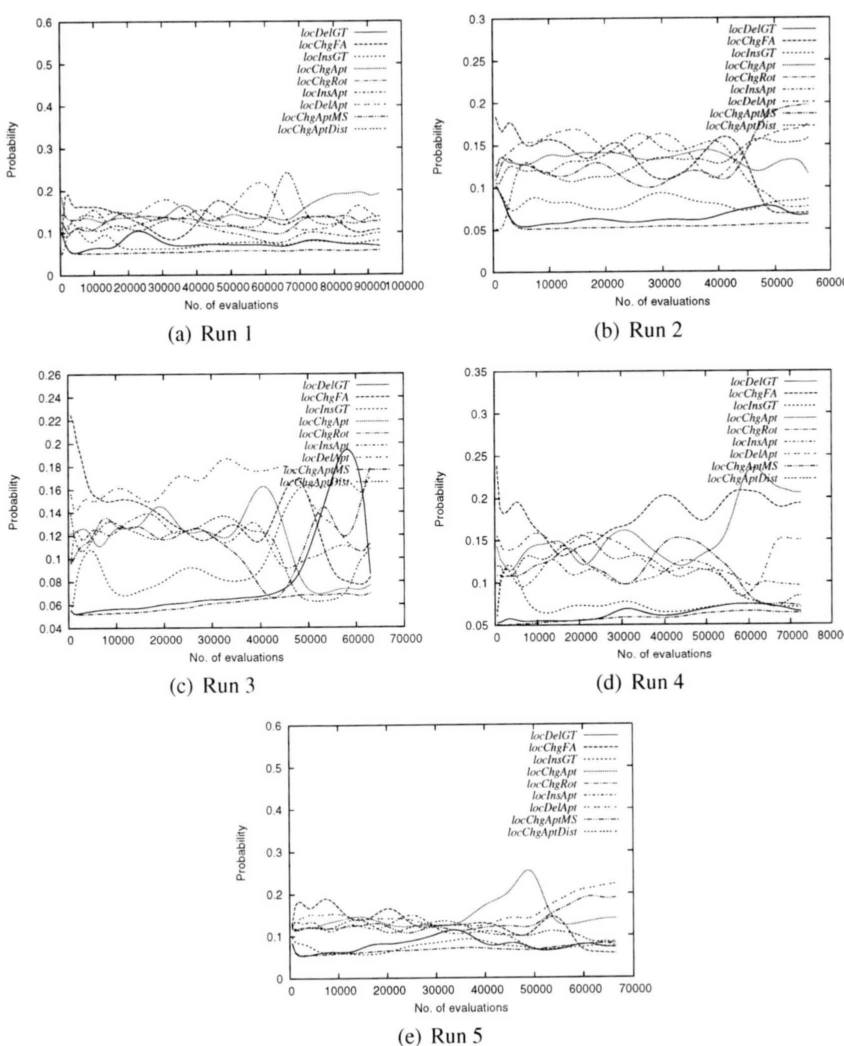

(a) Run 1

(b) Run 2

(c) Run 3

(d) Run 4

(e) Run 5

Fig. C.36 Application probability of the different variants of the local search operators (scenario C)

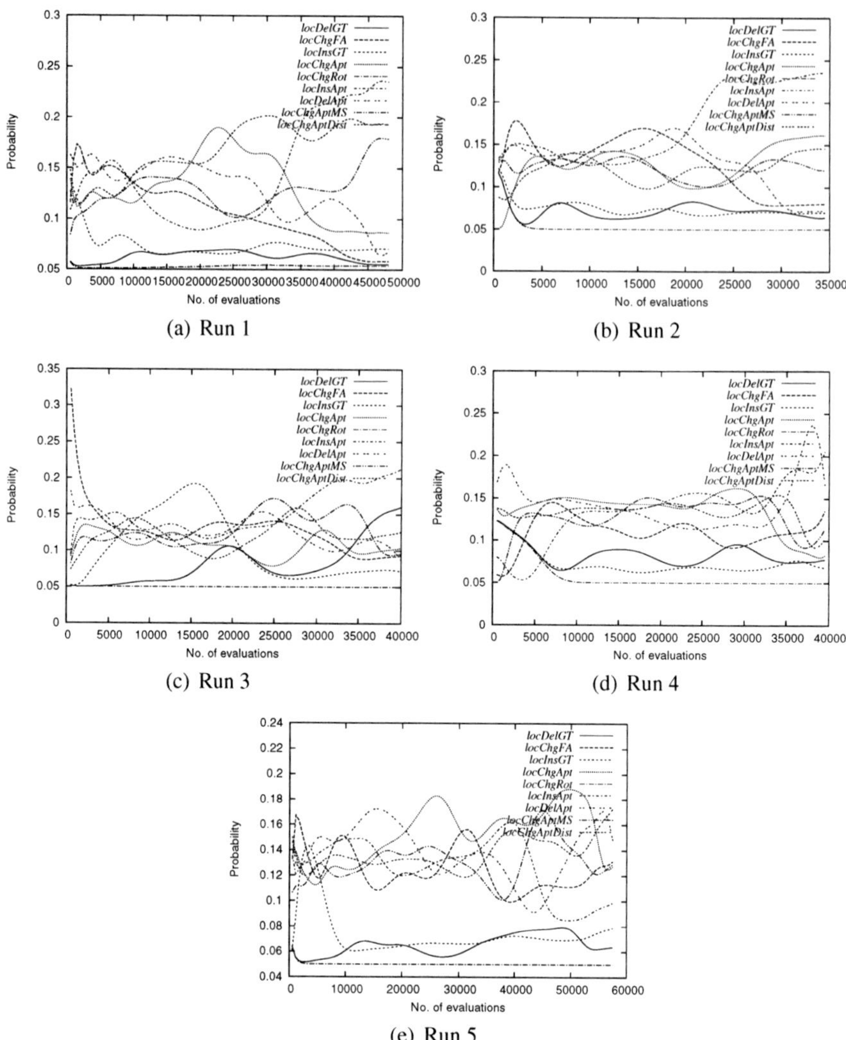

(a) Run 1

(b) Run 2

(c) Run 3

(d) Run 4

(e) Run 5

Fig. C.37 Application probability of the different variants of the local search operators (scenario D)

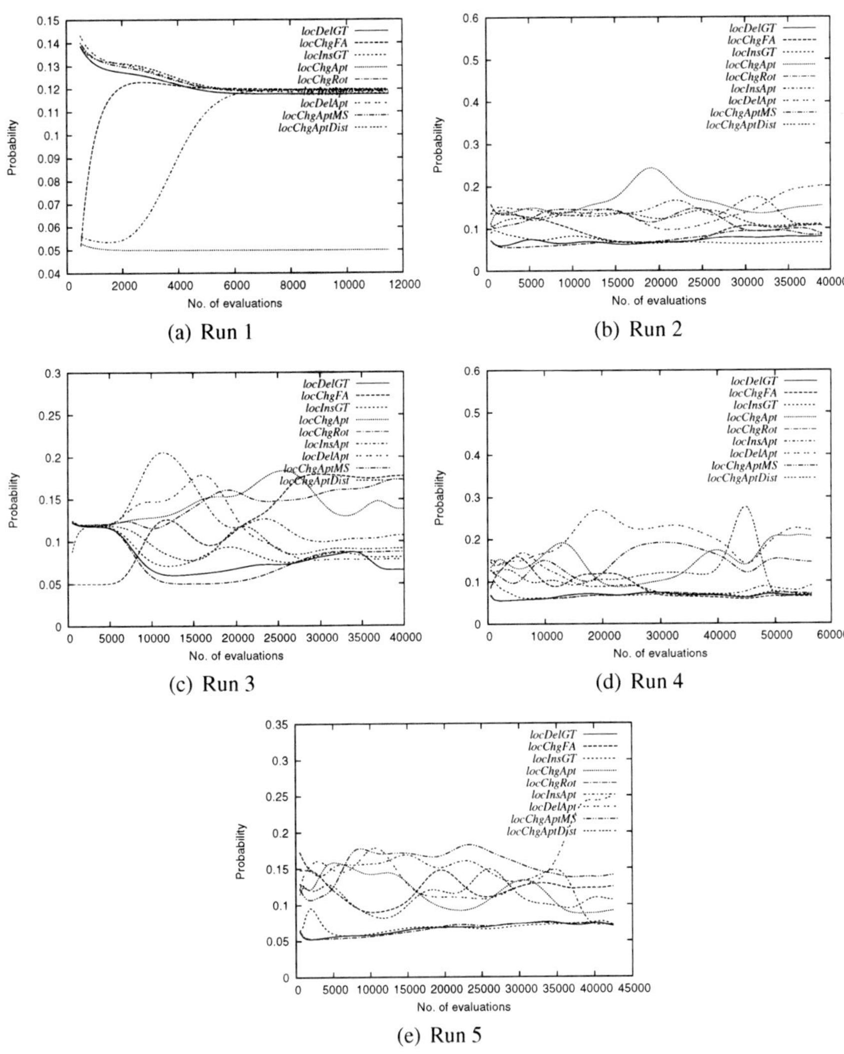

Fig. C.38 Application probability of the different variants of the local search operators (scenario E)

C.3 Evaluation

Table C.20 Profit of airline schedules constructed with the extended sequential planning approach

Run	Scenario				
	A	B	C	D	E
1	519,882	342,255	-9,872	90,675	109,782
2	504,730	337,968	-46,941	92,407	106,372
3	522,984	323,417	-57,198	89,527	115,100
4	510,847	316,572	-45,669	83,895	125,336
5	456,986	339,030	-43,546	79,368	111,804
Average	503,086	331,848	-40,645	87,174	113,679
Stand. dev.	26,768	11,200	17,988	5,405	7,246

Table C.21 Seat load factors (SLF) of airline schedules constructed with the extended sequential planning approach

Run	Scenario				
	A	B	C	D	E
1	0.224	0.401	0.188	0.409	0.290
2	0.292	0.400	0.183	0.428	0.299
3	0.298	0.366	0.195	0.392	0.278
4	0.309	0.365	0.194	0.374	0.298
5	0.293	0.400	0.190	0.391	0.282
Average	0.283	0.386	0.190	0.399	0.289
Stand. dev.	0.034	0.019	0.005	0.020	0.009

Table C.22 Numbers of passengers of airline schedules constructed with the extended sequential planning approach

Run	Scenario				
	A	B	C	D	E
1	4,802	3,882	1,872	2,156	2,125
2	5,666	3,643	1,818	2,359	2,142
3	5,803	3,876	1,855	2,298	1,991
4	6,366	3,661	1,930	2,216	2,267
5	6,002	3,587	1,693	2,186	1,901
Average	5,728	3,730	1,834	2,243	2,085
Stand. dev.	581	139	88	84	142

Table C.23 Numbers of flights of airline schedules constructed with the extended sequential planning approach

| | Scenario | | | | |
Run	A	B	C	D	E
1	103	103	60	69	49
2	128	95	66	73	47
3	130	107	62	77	46
4	140	107	61	79	51
5	133	97	58	71	40
Average	126.80	101.80	61.40	73.80	46.60
Stand. dev.	14.06	5.59	2.97	4.15	4.16

Table C.24 Numbers of fitness evaluations required by the extended sequential planning approach

| | Scenario | | | | |
Run	A	B	C	D	E
1	364,001	92,989	290,734	77,715	224,994
2	822,709	29,799	88,073	141,324	207,728
3	1,021,445	257,728	473,652	147,526	185,896
4	1,346,583	62,146	651,020	65,740	13,508
5	712,680	296,375	669,691	74,690	238,814
Average	853,484	147,807	434,634	101,399	174,188
Stand. dev.	364,597	120,855	247,189	39,584	91,975

References

Abara, J.: Applying integer linear programming to the fleet assignment problem. Interfaces 19(4), 20–28 (1989)

ACI (Airports Council International). Night flight restrictions. Report. Brussels (2004)

Ackhoff, R.: Science in the systems age: Beyond IE, OR and MS. Operations Research 21, 661–671 (1973)

Adler, N.: Competition in a deregulated air transportation market. European Journal of Operational Research 129, 337–345 (2001)

AEA (Association of European Airlines). 2005 summary of traffic and airline results. Brussels (2005)

Airbus. Global market forecast 2006 – 2025. Blagnac Cedex (2006)

Alamdari, F.E., Black, I.G.: Passengers' choice of airline under competition: The use of the logit model. Transport Reviews 12(2), 153–170 (1992)

Alander, J.T.: Indexed bibliography of genetic algorithms in economics. Technical Report 94-1-ECO. Department of Information Technology and Production Economics, University of Vaasa (2000)

Alderighi, M., Cento, A., Nijkamp, P., Rietveld, P.: Network competition – the coexistence of hub-and-spoke and point-to-point systems. Journal of Air Transport Management 11, 328–334 (2005)

Anbil, R., Forrest, J.J., Pulleyblank, W.R.: Column generation and the airline crew pairing problem. Documenta Mathematica, Extra volume ICM(III), 677–686 (1998)

Anbil, R., Gelman, E., Patty, B., Tanga, R.: Recent advances in crew pairing optimization at American Airlines. Interfaces 21, 62–74 (1991)

Anbil, R., Johnson, E., Tanga, R.: A global approach to crew pairing optimization. IBM Systems Journal 31, 71–78 (1992)

Andersson, E., Efthymios, H., Kohl, N., Wedelin, D.: Crew pairing optimization. In: Yu, G. (ed.) Operations Research in the Airline Industry, pp. 1–31. Kluwer Academic Publishers, Boston (1998)

Andersson, S.-E.: Operational planning in airline business – Can science improve efficiency? Experiences from SAS. European Journal of Operational Research 43, 3–12 (1989)

Antes, J.: Structuring the process of airline scheduling. In: Kischka, P. (ed.) Operations Research Proceedings 1997, Selected Papers of the Symposium on Operations Research (SOR 1997), pp. 515–520. Springer, Berlin (1998)

Antes, J., Campen, L., Derigs, U., Titze, C., Wolle, G.-D.: SYNOPSE: a model-based decision support system for the evaluation of flight schedules for cargo airlines. Decision Support Systems 22(4), 307–323 (1998)

Arabeyre, J., Fearnley, J., Steiger, F., Teather, W.: The airline crew scheduling problem: A survey. Transportation Science 3, 140–163 (1969)

Argüello, M.F., Bard, J.F., Yu, G.: A GRASP for aircraft routing in response to groundings and delays. Journal of Combinatorial Optimization 5, 211–228 (1997)

Argüello, M.F., Bard, J.F., Yu, G.: Models and methods for managing airline irregular operations aircraft routing. In: Yu, G. (ed.) Operations Research in the Airline Industry, pp. 1–45. Kluwer Academic Publishers, Boston (1998)

Armacost, A.P., Barnhart, C., Ware, K., Wilson, A.M.: UPS optimizes its air network. Interfaces 34(1), 15–25 (2004)

Armacost, A.P., Barnhart, C., Ware, K.A.: Composite variable formulations for express shipment service network design. Transportation Science 36(1), 1–20 (2002)

Ashford, N., Benchemam, M.: Passengers choice of airport: An application of the multinomial logit model. Transportation Research Record 1147, 1–5 (1987)

ATA (Air Transport Association of America). 2002 annual report. Washington (2002)

ATA (Air Transport Association of America). The airline handbook – online version (2003a)

ATA (Air Transport Association of America). Annual passenger prices (yield) (2003b) (March 22, 2003), http://www.airlines.org/public/industry/display1.asp?nid=1035

Aykin, T.: Networking policies for hub-and-spoke systems with application to the air transportation system. Transportation Science 29(3), 201–221 (1995)

Bäck, T., Fogel, D.B., Michalewicz, Z.: Handbook of Evolutionary Computation. Institute of Physics Publishing and Oxford University Press, Bristol and New York (1997)

Baker, E.K., Bodin, L.D., Finnegan, W.F., Ponder, R.: Efficient heuristic solutions to an airline crew scheduling problem. AIIE Transactions 11(2), 79–85 (1979)

Balakrishnan, A., Chien, T.W., Wong, R.T.: Selecting aircraft routes for long-haul operations: A formulation and solution method. Transportation Research Part B: Methodological 24(1), 57–72 (1990)

Ball, M., Roberts, A.: A graph partitioning approach to airline crew scheduling. Transportation Science 19, 107–126 (1985)

Ball, M.O.: Introduction to the special issue on aviation operations research: Commemorating 100 years of aviation. Transportation Science 37(4), 366–367 (2003)

Bania, N., Bauer, P.W., Zlatoper, T.J.: U.S. air passenger service: A taxonomy of route networks, hub locations, and competition. Transportation Research Part E: Logistics and Transportation Review 34(1), 53–74 (1998)

Bard, J., Kontoravdis, G., Yu, G.: A branch-and-cut procedure for the vehicle routing problem with time windows. Transportation Science 36(2), 250–269 (2002)

Bard, J.F., Cunningham, I.G.: Improving through-flight schedules. IIE Transportation 19(3), 242–251 (1987)

Barnett, A., Curtis, T., Goranson, J., Patrick, A.: Better than ever: Nonstop jet service in an era of hubs and spokes. Sloan Management Review 33(2), 49–54 (1992)

Barnhart, C., Belobaba, P.P., Odoni, A.R.: Applications of operations research in the air transport industry. Transportation Science 37(4), 368–391 (2003)

Barnhart, C., Boland, N.L., Clarke, L.W., Johnson, E.L., Nemhauser, G.L., Shenoi, R.G.: Flight string models for aircraft fleeting and routing. Transportation Science 32(3), 208–220 (1998)

Barnhart, C., Hatay, L., Johnson, E.L.: Deadhead selection for the long-haul crew pairing problem. Operations Research 43, 491–499 (1995)

Barnhart, C., Johnson, E.L., Anbil, R., Hatay, L.: A column generation technique for the long-haul crew assignment problem. In: Ciriani, T., Leachman, R. (eds.) Optimization in Industry, vol. 2, pp. 7–24. Wiley & Sons, Chichester (1994)

Barnhart, C., Johnson, E.L., Nemhauser, G.L., Savelsbergh, M.W.P., Vance, P.H.: Branch-and-price: Column generation for solving huge integer programs. Operations Research 46(3), 316–329 (1998)

Barnhart, C., Kniker, T.S., Lohatepanont, M.: Itinerary-based airline fleet assignment. Transportation Science 36(2), 199–217 (2002)

Barnhart, C., Lu, F., Shenoi, R.: Integrated airline scheduling planning. In: Yu, G. (ed.) Operations Research in the Airline Industry. International Series in Operations Research and Management Science, pp. 384–403. Kluwer Academic Publishers, Boston (1998)

Barnhart, C., Schneur, R.R.: Air network design for express shipment service. Operations Research 44(6), 852–863 (1996)

Barnhart, C., Shenoi, R.G.: An approximate model and solution approach for the long-haul crew pairing problem. Transportation Science 32(3), 221–231 (1998)

Barnhart, C., Talluri, K.T.: Airline operations research. In: ReVelle, C., McGarity, A.E. (eds.) Design and Operation of Civil and Environmental Engineering Systems, pp. 435–469. Wiley, New York (1997)

Barth, T.: Erzeugung und optimierung von vollständigen flugplänen auf basis des sequentiellen planungsprozesses. Diploma thesis, University of Mannheim (2005)

Barutt, J., Hull, T.: Airline crew scheduling: Supercomputers and algorithms. SIAM News 23(6), 20–22 (1990)

Bauer, C.U.: ökonomische bewertung von flugplänen im rahmen einer simultanen flugplanerstellung. Diploma thesis, University of Mannheim (2004)

Bélanger, N., Desaulniers, G., Soumis, F., Desrosiers, J.: Periodic airline fleet assignment with time windows, spacing constraints, and time dependent revenues. European Journal of Operational Research 175, 1754–1766 (2006)

Belobaba, P.P.: Airline yield management: An overview of seat inventory control. Transportation Science 21(2), 63–73 (1987)

Belobaba, P.P., Farkas, A.: Yield management impacts on airline spill estimation. Transportation Science 33(2), 217–232 (1999)

Ben-Akiva, M., Bierlaire, M.: Discrete choice methods and their application to short term travel decisions. In: Hall, R. (ed.) Handbook of Transportation Science, vol. 23, pp. 5–34. Kluwer, Dordrecht (1999)

Ben-Akiva, M., Lerman, S.R.: Discrete choice analysis: Theory and application to travel demand. MIT Press, Cambridge (1985)

Berechman, J., Shy, O.: The structure of airline equilibrium networks. In: Bergh, J., Nijkamp, P., Rietveld, P. (eds.) Recent Advances in Spatial Equilibrium Modelling, pp. 138–155. Springer, Berlin (1996)

Berge, M.E., Hopperstad, C.A.: Demand driven dispatch: A method for dynamic aircraft capacity assignment, models and algorithms. Operations Research 41(1), 153–168 (1993)

Biermann, T.: Die bedeutung der "sechsten freiheit" für den wettbewerb im luftverkehr, Buchreihe des Instituts für Verkehrswissenschaft an der Universität zu Köln, vol. 46. Cologne (1986)

Biethahn, J., Nissen, V.: Evolutionary algorithms in management applications. Springer, Berlin (1995)

Bihr, R.A.: A conceptual solution to the aircraft gate assignment problem using 0,1 linear programming. Computers and Industrial Engineering 19(1-4), 280–284 (1990)

de Billette, E.V.: Regulation in the air: price-and-frequency caps. Transportation Research Part E: Logistics and Transportation Review 40(6), 465–476 (2004)

Blum, C., Roli, A.: Metaheuristics in combinatorial optimization: Overview and conceptual comparison. ACM Computing Surveys 35(3), 268–308 (2003)

Bodin, L., Golden, B., Assad, A., Ball, M.: Routing and scheduling of vehicles and crews: the state of the art. Computers and Operations Research 10(2), 63–211 (1983)

Boeing: Current market outlook. Seattle (2006)

Bolat, A.: Assigning arriving flights at an airport to the available gates. Journal of the Operational Research Society 50(1), 23–34 (1999)

Bolat, A.: Procedures for providing robust gate assignments for arriving aircrafts. European Journal of Operational Research 120, 63–80 (2000)

Bondy, J.A., Murty, U.S.R.: Graph theory with applications. McMillan, London (1978)

Bonissone, P., Subbu, R., Eklund, N., Kiehl, T.: Evolutionary algorithms + domain knowledge = real-world evolutionary computation. IEEE Transactions on Communications 10(3), 256–280 (2006)

Borenstein, S., Netz, J.: Why do all the flights leave at 8 am?: Competition and departure-time differentiation in airline markets. International Journal of Industrial Organization 17, 611–640 (1999)

Bouamrene, M.A., Flavell, R.: Airline corporate planning – a conceptual framework. Long Range Planning 13(1), 62–69 (1980)

Brown, S., Watkins, W.S.: The demand for air travel. A regression study of time series and cross-sectional data in the U.S. domestic market. Highway Research Record (213), 21–34 (1968)

Brueckner, J.K., Zhang, Y.: A model of scheduling in airline networks: How a hub-and-spoke system affects flight frequency, fares and welfare. Journal of Transport Economics and Policy 35(2), 195–222 (2001)

BTS (Bureau of Transportation Statistics). Air carrier statistics (2006) (October 17, 2006), `http://www.transtats.bts.gov/`

Büdenbender, K., Grünert, T., Sebastian, H.-J.: A hybrid tabu search/branch-and-bound algorithm for the direct flight network design problem. Transportation Science 34(4), 364–380 (2000)

Butchers, E.R., Day, P.R., Goldie, A.P., Miller, S., Meyer, J.A., Ryan, D.M., Scott, A.C., Wallace, C.A.: Optimized crew scheduling at Air New Zealand. Interfaces 31(1), 30–56 (2001)

Button, K.: How stable are scheduled air transport markets. Research in Transportation Economics 13, 27–48 (2005)

Campbell, J.F.: Integer programming formulations of discrete hub location problems. European Journal of Operational Research 72, 387–405 (1994a)

Campbell, J.F.: A survey of network hub location. Studies in Location Analysis 6, 31–49 (1994b)

Campbell, K.W., Durfee, R.B., Hines, G.S.: FedEx generates bid lines using simulated annealing. Interfaces 27(2), 1–16 (1997)

Caprì, S., Ignaccolo, M.: Genetic algorithms for solving the aircraft-sequencing problem: the introduction of departures into the dynamic model. Journal of Air Transport Management 10(5), 345–351 (2004)

Caruana, R.A., Schaffer, J.D.: Representations and hidden bias: Gray vs. binary coding for genetic algorithms. In: Laird, L. (ed.) Proceedings of the Fifth International Workshop on Machine Learning, pp. 153–161. Morgan Kaufmann, San Mateo (1988)

Caulkins, J.P., Barnett, A., Larkey, P.D., Yuan, Y., Goranson, J.: The on-time machines: Some analyses of airline punctuality. Operations Research 41(4), 710–720 (1993)

Cerny, V.: A thermodynamical approach to the travelling salesman problem: an efficient simulation algorithm. Journal of Optimization Theory and Applications 45, 41–51 (1985)

Chang, S.C.: A new aircrew-scheduling model for short-haul routes. Journal of Air Transport Management 8(4), 249–260 (2002)

Cheng, Y.: Network-based simulation of aircraft at gates in airport terminals. Journal of Transportation Engineering, 188–196 (1998)

Chestler, L.: Overnight air express: Spatial pattern, competition and the future of small package delivery services. Transportation Quarterly 39, 59–71 (1985)

Christou, I.T., Zakarian, A., Liu, J.-M., Carter, H.: A two-phase genetic algorithm for large-scale bidline-generation problems at Delta Air Lines. Interfaces 29(5), 51–65 (1999)

Chu, H.D., Gelman, E., Johnson, E.L.: Solving large scale crew scheduling problems. European Journal of Operational Research 97, 260–268 (1997)

Clarke, L.W., Hane, C.A., Johnson, E.L., Nemhauser, G.L.: Maintenance and crew considerations in fleet assignment. Transportation Science 30(3), 249–260 (1996)

Clarke, L.W., Johnson, E., Nemhauser, G.L., Zhu, Z.: The aircraft rotation problem. Annals of Operations Research 69, 33–46 (1997)

Clarke, M.D.D.: Irregular airline operations: A review of the state-of-the-practice in airline operations control centers. Journal of Air Transport Management 4(2), 67–76 (1998)

Cohn, A.M., Barnhart, C.: Improving crew scheduling by incorporating key maintenance routing decisions. Operations Research 51(3), 387–396 (2003)

Coldren, G.M., Koppelman, F.S.: Modeling the competition among air-travel itinerary shares: GEV model development. Transportation Research Part A: Policy and Practice 39, 345–365 (2005)

Coldren, G.M., Koppelman, F.S., Kasturirangan, K., Mukherjee, A.: Modeling aggregate air-travel itinerary shares: Logit model development at a major US airline. Journal of Air Transport Management 9, 361–369 (2003)

Copeland, D.G., Mason, R.O., McKenney, J.L.: Sabre: The development of information-based competence and execution of information-based competition. IEEE Annals of the History of Computing 17(3), 30–55 (1995)

Cordeau, J.-F., Stojkovic, G., Soumis, F., Desrosiers, J.: Benders decomposition for simultaneous aircraft routing and crew scheduling. Transportation Science 35(4), 375–388 (2001)

Costaguta, A.A., Resiak, G.: Regional differences persist in operating economics of international airlines. ICAO Journal 57(6), 7–11 (2002)

Cote, J.-P., Marcotte, P., Savard, G.: A bilevel modelling approach to pricing and fare optimisation in the airline industry. Journal of Revenue and Pricing Management 2(1), 23–36 (2003)

Crainic, T.G.: Service network design in freight transportation. European Journal of Operational Research 122, 272–288 (2000)

Crainic, T.G., Rousseau, J.-M.: The column generation principle and the airline crew scheduling problem. INFOR 25, 136–151 (1987)

Crainic, T.G., Roy, J.: OR tools for tactical freight transportation planning. European Journal of Operational Research 33(3), 290–297 (1988)

Danesi, A.: Measuring airline hub timetable co-ordination and connectivity: definition of a new index and application to a sample of European hubs. European Transport / Transporti Europei 34, 54–74 (2006)

Dantzig, G.B.: Linear programming and extensions. Princeton University Press, Princeton (1963)

Darwin, C.: On the origin of species. John Murray, London (1859)

Daskin, M.S., Panayotopoulos, N.D.: A langrangian relaxation approach to assigning aircraft to routes in hub and spoke networks. Transportation Science 23(2), 91–99 (1989)

Davis: Handbook of genetic algorithms. Van Nostrand Reinhold, New York (1991)

Dawid, H., König, J., Strauss, C.: An enhanced rostering model for airline crews. Computers and Operations Research 28, 671–688 (2001)

Day, P.R., Ryan, D.M.: Flight attendant rostering for short-haul airline operations. Operations Research 45(5), 649–661 (1997)

De Jong, K.A.: An analysis of the behavior of a class of genetic adaptive systems. Dissertation, University of Michigan (1975)

De Jong, K.A., Sarma, J.: Generation gaps revisited. In: Whitley, D. (ed.) Foundations of Genetic Algorithms, 2nd edn., pp. 19–28. Morgan Kaufmann, San Mateo (1993)

Dennis, N.: Airline hub operations in Europe. Journal of Transport Geography 2(4), 219–233 (1994a)

Dennis, N.: Scheduling strategies for airline hub operations. Journal of Air Transport Management 1(3), 131–144 (1994b)

Dennis, N.: Scheduling issues and network strategies for international airline alliances. Journal of Air Transport Management 6(2), 75–85 (2000)

Desaulniers, G., Desrosiers, J., Dumas, Y., Marc, S., Rioux, B., Solomon, M.M., Soumis, F.: Crew pairing at Air France. European Journal of Operational Research 97(2), 245–259 (1997)

Desaulniers, G., Desrosiers, J., Dumas, Y., Solomon, M.M., Soumis, F.: Daily aircraft routing and scheduling. Management Science 43(6), 841–855 (1997)

Desaulniers, G., Desrosiers, J., Gamache, M., Soumis, F.: Crew scheduling in air transportation. In: Crainic, T., Laporte, G. (eds.) Volume on Network Routing, Fleet Management and Logistics, pp. 169–185. Kluwer, Norwell (1998)

Desaulniers, G., Desrosiers, J., Ioachim, I., Soumis, F., Solomon, M.: A unified framework for time constrained vehicle routing and crew scheduling problems. In: Crainic, T., Laporte, G. (eds.) Fleet Management and Logistics, pp. 67–93. Kluwer, Norwell (1998)

Desaulniers, G., Desrosiers, J., Lasry, A., Solomin, M.M.: Crew pairing for a regional carrier. In: Wilson, N.H.M. (ed.) Computer-Aided Transit Scheduling. Lecture Notes in Economics and Mathematical Systems, vol. 471, pp. 19–41. Springer, Berlin (1997)

Desrochers, M., Lenstra, J.K., Savelsbergh, M.W.P.: A classification theme for vehicle routing and scheduling problems. European Journal of Operational Research 46, 322–332 (1990)

Desrosiers, J., Dumas, Y., Solomon, M., Soumis, F.: Time constrained routing and scheduling. In: Ball, M., Magnanti, C.M., Nemhauser, G.L. (eds.) Handbook in Operations Research/Management Science, Network Routing, vol. 8, pp. 35–139. Elsevier Science Publishers B.V, Amsterdam (1995)

Desrosiers, J., Lasry, A., McInnis, D., Solomon, M.M., Soumis, F.: Air Transat uses ALTITUDE to manage its aircraft routing, crew pairing, and work assignment. Interfaces 30(2), 41–53 (2000)

Dillon, J.E., Kontogiorgis, S.: US Airways optimizes the scheduling of reserve flight crews. Interfaces 29(5), 123–131 (1999)

Dobson, G., Lederer, P.J.: Airline scheduling and routing in a hub-and-spoke system. Transportation Science 27(3), 281–297 (1993)

Doganis, R.: Traffic forecasting and the gravity model. Flight International 29, 547–549 (1966)

Doganis, R.: Flying off course – the economics of international airlines, 3rd edn. Routledge, London (2004)

Domschke, W., Drexl, A.: Einführung in operations research, 6th edn. Springer, Berlin (2005)

Douglas, G.W., Miller, J.C.: Economic regulation of domestic air transport: Theory and policy. The Brookings Institution (1974)

Droste, S., Wiesmann, D.: On the design of problem-specific evolutionary algorithms. In: Ghosh, A., Tsutsui, S. (eds.) Advances in Evolutionary Computing: Theory and Applications, pp. 153–173. Springer, Heidelberg (2002)

Dueck, Scheuer: Threshold accepting: A general purpose optimization algorithm appearing superior to simulated annealing. Journal of Computational Physics 90(1), 161–175 (1990)

Duffuaa, S.O., Andijani, A.A.: Integrated simulation model for effective planning of maintenance operations for saudi arabian airlines (saudia). Production Planning and Control 10(6), 579–584 (1999)

EC (European Commission). Assessing the economic costs of night flight restrictions. Final Report Tender No. TREN/F3/10-2003. London (2005)

Emden-Weinert, T., Proksch, M.: Best practice simulated annealing for the airline crew scheduling problem. Journal of Heuristics 5(4), 419–436 (1999)

Erdmann, A., Nolte, A., Noltemeier, A., Schrader, R.: Modeling and solving an airline schedule generation problem. Annals of Operations Research 107(1-4), 117–142 (2001)

Etschmaier, M.M., Mathaisel, D.F.X.: Airline scheduling: An overview. Transportation Science 19(2), 127–138 (1985)

Feo, T.A., Bard, J.F.: Flight scheduling and maintenance base planning. Management Science 35(12), 1415–1432 (1989)

Ferguson, A.R., Dantzig, G.B.: The allocation of aircraft to routes – an example of linear programming under uncertain demand. Management Science 3, 45–73 (1956a)

Ferguson, A.R., Dantzig, G.B.: The problem of routing aircraft. Aeronautival Engineering Review 14 (1956b)

Filar, J.A., Manyem, P., White, K.: How airlines and airports recover from schedule perturbations: A survey. Annals of Operations Research 108(1-4), 315–333 (2001)

Fotheringham, A.S.: Some theoretical aspects of destination choice and their relevance to production-constrained gravity models. Environment and Planning A 15, 1121–1132 (1983)

Fotheringham, A.S., Webber, M.J.: Spatial structure and the parameters of spatial interaction models. Geographical Analysis 12(1), 33–46 (1980)

Franken, R.: Objektorientierte Gestaltung von Planungsunterstützungssystemen für die Produktionsplanung – Dargestellt am Beispiel der Kapazitäts- und Flugplanung der Deutschen Lufthansa AG. Wirtschaftsinformatik 32(3), 253–262 (1990)

Freling, R., Wagelmans, A.P.M., Pinto Paixao, J.M.: An overview of models and techniques for integrating vehicle and crew scheduling. In: Wilson, N.H.M. (ed.) Computer-Aided Transit Scheduling. Lecture Notes in Economics and Mathematical Systems, vol. 471, pp. 441–460. Springer, Berlin (1997)

Gamache, M., Soumis, F.: A method for optimally solving the rostering problem. In: Yu, G. (ed.) Operations Research in the Airline Industry, pp. 124–157. Kluwer Academic Publishers, Boston (1998)

Gamache, M., Soumis, F., Marquis, G., Desrosiers, J.: A column generation approach for large-scale aircrew rostering problems. Operations Research 47(2), 247–262 (1999)

Gamache, M., Soumis, F., Villeneuve, D., Desrosiers, J., Gelinas, E.: The preferential bidding system at air canada. Transportation Science 32(3), 246–255 (1998)

Gardner Jr., E.S.: Dimensional analysis of airline quality. Interfaces 34(4), 272–279 (2004)

Gelerman, W., de Neufville, R.: Planning for satellite airports. Journal of the Transportation Engineering Division 99(3), 537–551 (1973)

Gendron, B., Crainic, T.G., Frangioni, A.: Multicommodity capacitated network design. In: Sanso, B., Soriano, P. (eds.) Telecommunications Network Planning, pp. 1–19. Kluwer Academic Publishers, Dordrecht (1999)

Gershkoff, I.: Optimizing flight crew schedules. Interfaces 19(4), 29–43 (1989)

Gertsbach, I., Gurevich, Y.: Constructing an optimal fleet for a transportation schedule. Transportation Science 11(1), 20–36 (1977)

Ghobrial, A., Balakrishnan, N., Kanafani, A.: A heuristic model for frequency planning and aircraft routing in small size airlines. Transportation Planning and Technology 16(4), 235–249 (1992)

Ghobrial, A., Kanafani, A.: Future of airline hubbed networks: Some policy implications. Journal of Transportation Engineering 121(2), 124–134 (1995a)

Ghobrial, A.A., Kanafani, A.: Quality-of-service model of intercity air-travel demand. Journal of Transportation Engineering 121(2), 135–140 (1995b)

Glover, F.: Future paths for integer programming and links to artificial intelligence. Computers and Operations Research 13(5), 533–549 (1986)

Glover, F., Kochenberger, G.A.: Handbook of metaheuristics. Kluwer, Boston (2003)

Goldberg, D.E.: Genetic algorithms in search, optimization and machine learning. Addison-Wesley, Reading (1989)

Goldberg, D.E.: The design of innovation. Series on Genetic Algorithms and Evolutionary Computation. Kluwer, Dordrecht (2002)

Goldberg, D.E., Korb, B., Deb, K.: Messy genetic algorithms: Motivation, analysis, and first results. Complex Systems 3(5), 493–530 (1989)

Golden, B.L., Assad, A.A.: Vehicle routing: Methods and studies. Elsevier, Amsterdam (1988)

Golden, B.L., Wong, R.T.: Vehicle routing by land, sea, and air. Interfaces 22(3), 1–3 (1992)

Gopalan, R., Talluri, K.T.: The aircraft maintenance routing problem. Operations Research 46(2), 260–271 (1998a)

Gopalan, R., Talluri, K.T.: Mathematical models in airline schedule planning: A survey. Annals of Operations Research 76, 155–185 (1998b)

Gosling, G.: Design of an expert system for aircraft gate assignment. Transportation Research Part A: Policy and Practice 24(1), 59–69 (1990)

Grandeau, S.C., Clarke, M.D., Mathaisel, D.F.X.: The processes of airline system operations control. In: Yu, G. (ed.) Operations Research in the Airline Industry, pp. 312–369. Kluwer Academic Publishers, Boston (1998)

Graves, G.W., Mcbride, R.D., Gershkoff, I., Anderson, D., Mahidhara, D.: Flight crew scheduling. Management Science 39(6), 736–745 (1993)

Grefenstette, J.J.: Proceedings of an international conference on genetic algorithms and their applications. Lawrence Erlbaum Associates, Hillsdale (1985)

Grosche, T., Rothlauf, F., Heinzl, A.: Gravity models for airline passenger volume estimation. Journal of Air Transport Management (2007)

Gu, Y., Chung, C.A.: Genetic algorithm approach to aircraft gate reassignment problem. Journal of Transportation Engineering 125(5), 384–389 (1999)

Gu, Z., Johnson, E.L., Nemhauser, G.L., Wang, Y.: Some properties of the fleet assignment problem. Operations Research Letters 15, 59–71 (1994)

Guo, Y., Mellouli, T., Suhl, L., Thiel, M.P.: A partially integrated airline crew scheduling approach with time-dependent crew capacities and multiple home bases. European Journal of Operational Research 171, 1169–1181 (2006)

Gursoy, D., Chen, M.-H., Kim, H.J.: The US airlines relative positioning based on attributes of service quality. Tourism Management 26(1), 57–67 (2003)

Haghani, A., Chen, M.-C.: Optimizing gate assignments at airport terminals. Transportation Research Part A: Policy and Practice 32(6), 437–454 (1998)

Hall, R.: Configuration of an overnight package air network. Transportation Research 23(2), 139–149 (1989)

Hane, C.A., Barnhart, C., Sigismindi, G.: The fleet assignment problem: Solving a large-scale integer program. The Mathematical Programming Society 70(4), 211–232 (1995)

Hansen, M.: Airline competition in a hub-dominated environment: An application of non-cooperative game theory. Transportation Research Part B: Methodological 24(1), 27–43 (1990)

Hillier, F.S., Lieberman, G.J.: Introduction to operations research, 7th edn. McGraw-Hill, New York (2002)

Hochfeld, C., Arps, H., Hermann, A., Otten, S., Schmied, M.: Untersuchungen an internationalen und nationalen Verkehrsflughäfen zum Mediationspaket – State-Of-Practice-Analyse –. Final Report Öko-Institut e.V. Darmstadt/Berlin (2003)

Hoffman, K.L., Padberg, M.: Solving airline crew-scheduling problems by branch-and-cut. Management Science 39(6), 657–682 (1993)

Holland, J.H.: Adaption in natural and artificial systems. University of Michigan Press, Ann Arbor (1975)

Hsu, C.-I., Wen, Y.-H.: Application of grey theory and multiobjective programming towards airline network design. European Journal of Operational Research 127, 44–68 (2000)

Hsu, C.-I., Wen, Y.-H.: Determining flight frequencies on an airline network with demand-supply interactions. Transportation Research Part E: Logistics and Transportation Review 39(6), 417–442 (2003)

Hsu, C.-I., Wu, Y.-H.: The market size of a city-pair route at an airport. The Annals of Regional Science 31, 391–409 (1997)

ICAO (International Civil Aviation Organization). Annual review of civil aviation 2005. ICAO Journal 61(5) (2006)

Ioachim, I., Desrosiers, J., Soumis, F., Blanger, N.: Fleet assignment and routing with schedule synchronization constraints. European Journal of Operational Research 119, 75–90 (1999)

Ippolito, R.A.: Estimating airline demand with quality of service variables. Journal of Transport Economics and Policy 15(1), 7–15 (1981)

Jacobs, T.L., Ratliff, R.M., Smith, B.C.: Soaring with synchronized systems. Coordinated scheduling, yield management and pricing decisions can make airline revenue take off. OR/MS Today 27(4), 36–44 (2000)

Jaillet, P., Song, G., Yu, G.: Airline network design and hub location problems. Location Science 4(3), 195–212 (1996)

Jarrah, A.I., Goodstein, J., Narasimhan, R.: An efficient airline re-fleeting model for the incremental modification of planned fleet assignments. Transportation Science 34(4), 349–363 (2000)

Jarrah, A.I.Z., Bard, J.F., de Silva, A.: Solving large-scale tour scheduling problems. Management Science 40(9), 1124–1144 (1994)

Jarrah, A.I.Z., Diamond, J.T.: The problem of generating crew bidlines. Interfaces 27(4), 49–64 (1997)

Jeng, C.-Y.: An idealized model for understanding impacts of key network parameters on airline routing. Transportation Research Record 1158, 5–13 (1988)

Jones, R.D.: Development of an automated airline crew bid generation system. Interfaces 19(4), 44–51 (1989)

Jorge-Calderón, J.D.: A demand model for scheduled airline services on international European routes. Journal of Air Transport Management 3(1), 23–35 (1997)

Kanafani, A.: Aircraft technology and network structure in short-haul air transportation. Transportation Research Part A: Policy and Practice 15, 305–314 (1981)

Kanafani, A.: Transportation demand analysis. McGraw-Hill Book Company, New York (1983)

Kasilingam, R.: Air cargo revenue management: Characteristics and complexities. European Journal of Operational Research, 36–44 (1997a)

Kasilingam, R.G.: An economic model for air cargo overbooking under stochastic capacity. Computers and Industrial Engineering 32(1), 221–226 (1997b)

Kim, D., Barnhart, C.: Transportation service network design: Models and algorithms. In: Wilson, N.H.M. (ed.) Computer-Aided Transit Scheduling. Lecture Notes in Economics and Mathematical Systems, vol. 471, pp. 259–283. Springer, Berlin (1997)

Kim, D., Barnhart, C., Ware, K., Reinhardt, G.: Multimodal express package delivery: A service network design application. Transportation Science 33(4), 391–407 (1999)

Kimes, S.E.: Yield management: A tool for capacity-constraint service firms. Journal of Operations Management 8(4), 348–363 (1989)

Kirkpatrick, S., Gelatt, C.D., Vecchi, M.P.: Optimization by simulated annealing. Science 220(4598), 671–680 (1983)

Klabjan, D., Johnson, E.L., Nemhauser, G.L., Gelman, E., Ramaswamy, S.: Airline crew scheduling with regularity. Transportation Science 35(4), 359–374 (2001a)

Klabjan, D., Johnson, E.L., Nemhauser, G.L., Gelman, E., Ramaswamy, S.: Solving large airline crew scheduling problems: Random pairing generation and strong branching. Computational Optimization and Applications 20(1), 73–91 (2001b)

Klabjan, D., Johnson, E.L., Nemhauser, G.L., Gelman, E., Ramaswamy, S.: Airline crew scheduling with time windows and plane-count constraints. Transportation Science 36(3), 337–348 (2002)

Kliewer, G.: Integrating market modeling and fleet assignment. In: Proceedings of the 2nd international Workshop CP-AI-OR 2000, pp. 89–92 (2000)

Kliewer, G., Tschöke, S.: A general parallel simulated annealing library and its application in airline industry. In: Proceedings of the 14th International Parallel and Distributed Processing Symposium (IPDPS 2000), pp. 55–61 (2000)

Klincewicz, J.G., Rosenwein, M.B.: The airline exception scheduling problem. Transportation Science 29(1) (1995)

Kontogiorgis, S., Acharya, S.: US Airways automates its weekend fleet assignment. Interfaces 29(3), 52–62 (1999)

Koza, J.R.: Genetic programming: On the programming of computers by natural selection. MIT Press, Cambridge (1992)

Kuby, M.J., Gray, R.G.: The hub network design problem with stopovers and feeders: The case of Federal Express. Transportation Research Part A: Policy and Practice 27(1), 1–12 (1993)

Kwok, L.-F., Hung, S.-L., Pun, C.-C.J.: Knowledge-based cabin crew pattern generator. Knowledge-Based Systems 8(1), 55–63 (1995)

Lagerholm, M., Peterson, C., Söderberg, B.: Airline crew scheduling using potts mean field techniques. European Journal of Operational Research 120, 81–96 (2000)

Lan, S., Clarke, J.-P., Barnhart, C.: Planning for robust airline operations: Optimizing aircraft routings and flight departure times to minimize passenger disruptions. Transportation Science 40(1), 15–28 (2006)

Langerman, J.J., Ehlers, E.M.: Agent-based airline scheduling. Computers and Industrial Engineering 33(3-4), 849–852 (1997)

Lavoie, S., Minoux, M., Odier, E.: A new approach for crew pairing problems by column generation with an application to air transportation. European Journal of Operational Research 35, 45–58 (1988)

Lederer, P.J.: A competitive network design problem with pricing. Transportation Science 27, 25–38 (1993)

Lederer, P.J., Nambimadom, R.S.: Airline network design. Operations Research 46(6), 785–804 (1998)

Lee, D.: Concentration and price trends in the US domestic airline industry: 1990-2000. Journal of Air Transport Management 9(2), 91–101 (2003)

Lee, L.H., Lee, C.U., Tan, Y.P.: A multi-objective genetic algorithm for robust flight scheduling using simulation. European Journal of Operational Research 177, 1948–1968 (2007)

Leibold, K.: Optimierung von flugplänen – anwendung quantitativer methoden im luftverkehr. Gabler, Wiesbaden (2001)

Lettovsky, L., Johnson, E.L., Nemhauser, G.L.: Airline crew recovery. Transportation Science 34(4), 337–348 (2000)

Levin, A.: Scheduling and fleet routing models for transportation systems. Transportation Science 5, 232–255 (1971)

Levine, D.: Application of a hybrid genetic algorithm to airline crew scheduling. Computers and Operations Research 23(6), 547–558 (1996)

Liepins, G.E., Vose, M.D.: Representational issues in genetic optimization. Journal of Experimental and Theoretical Artificial Intelligence 2, 101–115 (1990)

Link, M.: ökonomische bewertung von flugplänen im kombinierten passagier- und frachtflugverkehr. Diploma thesis, University of Mannheim (2006)

Lohatepanont, M., Barnhart, C.: Airline schedule planning: Integrated models and algorithms for schedule design and fleet assignment. Transportation Science 38(1), 19–32 (2004)

Lohmann, R.: Structure evolution and incomplete induction. Biological Cybernetics 69(4), 319–326 (1993)

Lucic, P., Teodorovic, D.: Simulated annealing for the multi-objective aircrew rostering problem. Transportation Research Part A: Policy and Practice 33, 19–45 (1999)

Lufthansa: Lufthansa timetable (April 17, 2007),
```
http://www.lufthansa.com/online/portal/lh/de/
info_and_services/flightinfo?nodeid=1770449&l=en
```

Luo, S., Yu, G.: On the airline schedule pertubation problem caused by the ground delay problem. Transportation Science 31(4), 298–311 (1997)

Luo, S., Yu, G.: Airline schedule perturbation problem: Ground delay program with splitable resources. In: Yu, G. (ed.) Operations Research in the Airline Industry, pp. 433–460. Kluwer Academic Publishers, Boston (1998a)

Luo, S., Yu, G.: Airline schedule perturbation problem: Landing and takeoff with nonsplitable resource for the ground delay program. In: Yu, G. (ed.) Operations Research in the Airline Industry, pp. 404–432. Kluwer Academic Publishers, Boston (1998b)

Magnanti, T.L., Wong, R.: Network design and transportation planning: Models and algorithms. Transportation Science 18(1), 1–55 (1984)

Manderick, B., de Weger, M., Spiessens, P.: The genetic algorithm and the structure of the fitness landscape. In: Belew, R.K., Booker, L.B. (eds.) Proceedings of the Fourth International Conference on Genetic Algorithms, pp. 143–150. Morgan Kaufmann, San Mateo (1991)

Mangoubi, R.S., Mathaisel, D.F.X.: Optimizing gate assignments at airport terminals. Transportation Science 19, 173–188 (1985)

Marianov, V., Serra, D., ReVelle, C.: Location of hubs in a competitive environment. European Journal of Operational Research 114, 363–371 (1999)

Marsten, R., Shepardson, F.: Exact solution of crew scheduling problems using the set partitioning model: Recent successful applications. Networks 11, 165–177 (1981)

Marsten, R.E., Muller, M.R.: A mixed-integer programming approach to air cargo fleet planning. Management Science 26, 1096–1107 (1980)

Mashford, J.S., Marksjo, B.S.: Airline base schedule optimisation by flight network annealing. Annals of Operations Research 108(1-4), 293–313 (2001)

Mason, R.O.: Absolutely, positively Operations Research: The Federal Express story. Interfaces 27(2), 17–36 (1997)

Mathaisel, D.F.X.: Decision support for airline schedule planning. Journal of Combinatorial Optimization 1, 251–275 (1997)

McFadden, D.: Conditional logit analysis of qualitative choice behavior. In: Zarembka, P. (ed.) Frontiers in Economics, pp. 105–142. Academic Press, New York (1974)

McGill, J.I., Van Ryzin, G.J.: Revenue management: Research overview and prospects. Transportation Science 33(2), 233–256 (1999)

Mendel, G.: Versuche über Pflanzenhybriden. Verhandlungen des naturforschenden Vereins 4, 3–47 (1866)

Michalewicz, Z., Fogel, D.B.: Handling constraints in genetic algorithm. In: Schaffer, J.D. (ed.) Proceedings of the Third International Conference on Genetic Algorithms, pp. 151–157. Morgan Kaufmann, San Mateo (1989)

Michalewicz, Z., Fogel, D.B.: How to solve it: Modern heuristics. Springer, Berlin (2000)

Michalewicz, Z., Schoenauer, M.: Evolutionary computation for constrained parameter optimization problems. Evolutionary Computation 4(1), 1–32 (1996)

Miller, R.: An optimization model for transportation planning. Transportation Research 1, 271–286 (1967)

Mingozzi, A., Boschetti, M.A., Ricciarde, S., Bianco, L.: A set partitioning approach to the crew scheduling problem. Operations Research 47(6), 873–888 (1999)

Minoux, M.: Network synthesis and optimum network design problems: Models, solution methods and applications. Networks 19, 313–360 (1989)

Moore, O.E., Soliman, A.H.: Airport catchment areas and air passenger demand. Transportation Engineering Journal of ASCE 107(TE5), 569–579 (1981)

Morrell, P.S., Pilon, R.V.: KLM and Northwest: a survey of the impact of a passenger alliance on cargo service characteristics. Journal of Air Transport Management 5, 153–160 (1999)

Morrison, S.A., Winston, C.: An economic analysis of the demand for intercity passenger transportation. Research in Transportation Economics 2, 213–237 (1985)

Moudani, W.E., Mora-Camino, F.: A dynamic approach for aircraft assignment and maintenance scheduling by airlines. Journal of Air Transport Management 6(4), 233–237 (2000)

Newton, I.: Philosophiae naturalis principia mathematica, London (1687)

Nicoletti, B.: Automatic crew rostering. Transportation Science 9, 33–42 (1975)

OAG (Official Airline Guide) About oag. (April 4, 2007), http://www.oag.com/

O'Connor, W.E.: An introduction to airline economics, 3rd edn. Praeger, New York (1982)

O'Kelly, M.E.: Activity levels at hub facilities in interacting networks. Geographical Analysis 18(4), 343–356 (1986)

O'Kelly, M.E.: A quadratic integer program for the location of interacting hub facilities. European Journal of Operational Research 32, 393–404 (1987)

O'Kelly, M.E., Song, W., Shen, G.: New estimates of gravitational attraction by linear programming. Geographical Analysis 27(4), 271–285 (1995)

Osman, I.H., Laporte, G.: Metaheuristics: A bibliography. Annals of Operations Research 63, 513–623 (1996)

Oum, T.-H., Tretheway, M.W.: Airline hub and spoke systems. Journal of Transportation Research Forum 30, 380–393 (1990)

Ozdemir, H.T., Mohan, C.K.: Graga: A graph based genetic algorithm for airline crew scheduling. In: Proceedings of the International Conference on Tools with Artificial Intelligence, pp. 27–28 (1999)

Ozdemir, H.T., Mohan, C.K.: Flight graph based genetic algorithm for crew scheduling in airlines. Information Sciences 133, 165–173 (2001)

Pak, K., Piersma, N.: Overview of OR techniques for airline revenue management. Statistica Neerlandica 56(4), 479–495 (2002)

Park, J.-W., Robertson, R., Wu, C.-L.: The effect of airline service quality on passengers' behavioural intentions: a Korean case study. Journal of Air Transport Management 10(6), 435–439 (2004)

Patty, B.W., Diamond, J.: The complex configuration model. In: Yu, G. (ed.) Operations Research in the Airline Industry, pp. 370–403. Kluwer Academic Publishers, Boston (1998)

Pearl, J.: Heuristics: Intelligent search strategies for computer problem solving. Addison-Wesley, Reading (1984)

Pollack, M.: Some aspects of the aircraft scheduling problem. Transportation Research 8, 233–243 (1974)

Proussaloglou, K., Koppelman, F.S.: The choice of air carrier, flight, and fare class. Journal of Air Transport Management 5(4), 193–201 (1999)

Puchta, M., Gottlieb, J.: Solving car sequencing problems by local optimization. In: Cagnoni, S., Gottlieb, J., Hart, E., Middendorf, M., Raidl, G.R. (eds.) EvoIASP 2002, EvoWorkshops 2002, EvoSTIM 2002, EvoCOP 2002, and EvoPlan 2002. LNCS, vol. 2279, pp. 131–140. Springer, Heidelberg (2002)

Pulugurtha, S.S., Nambisan, S.S.: Using genetic algorithms to evaluate aircraft ground holding policy in real time. Journal of Transportation Engineering 127(5), 442–449 (2001a)

Pulugurtha, S.S., Nambisan, S.S.: Using genetic algorithms to evaluate aircraft ground holding policy under static conditions. Journal of Transportation Engineering 127(5), 433–441 (2001b)

Pulugurtha, S.S., Nambisan, S.S.: A decision-support tool for airline yield management using genetic algorithms. Computer-Aided Civil and Infrastructure Engineering 18(3), 214–223 (2003)

Radcliffe, N.J.: Equivalence class analysis of genetic algorithm. Complex Systems 5(2), 183–205 (1991)

Radcliffe, N.J.: The algebra of genetic algorithms. Annals of Mathematics and Artificial Intelligence 10, 339–384 (1994)

Raguraman, K.: International air cargo hubbing: the case of Singapore. Asia Pacific Viewpoint 38(1), 55–74 (1997)

Rayward-Smith, V.J.: Applications of modern heuristic methods. Nelson Thornes Ltd. (1998)

Rayward-Smith, V.J., Osman, I.H., Reeves, C.R., Smith, G.D.: Modern heuristic search methods. Wiley, Chichester (1996)

Rechenberg, I.: Bionik, Evolution und Optimierung. Naturwissenschaftliche Rundschau 26(11), 465–472 (1973a)

Rechenberg, I.: Evolutionsstrategie: Optimierung technischer systeme nach prinzipien der biologischen evolution. Stuttgart - Bad Cannstatt: Friedrich Frommann Verlag (1973b)

Reeves, C.R.: Modern heuristic techniques for combinatorial problems. Blackwell Scientific Publications, Oxford (1993)

Reeves, C.R., Rowe, J.E.: Genetic algorithms: Principles and perspectives. Kluwer Academic Publishers, Dordrech (2003)

Rengaraju, V.R., Thamizh Arasan, V.: Modeling for air travel demand. Journal of Transportation Engineering 118(3), 371–380 (1992)

Rexing, B., Barnhart, C., Kniker, T., Jarrah, A., Krishnamurthy, N.: Airline fleet assignment with time windows. Transportation Science 34(1), 1–20 (2000)

Reynolds-Feighan, A.: Traffic distribution in low-cost and full-service carrier networks in the US air transportation market. Journal of Air Transport Management 7(5), 265–275 (2001)

Richardson, R.J.: An optimization approach to routing aircraft. Transportation Science 10, 52–71 (1976)

Richter, H.: Thirty years of airline Operations Research. Interfaces 19(4) (1989)

Rosenberger, J.M., Johnson, E.L., Nemhauser, G.L.: Rerouting aircraft for airline recovery. Transportation Science 37(4), 408–421 (2003)

Rothlauf, F.: Design and application of metaheuristics. Habilitationsschrift, University of Mannheim (2006a)

Rothlauf, F.: Representations fo genetic and evolutionary algorithms, 2nd edn. Springer, Heidelberg (2006b)

Rudolph, G., Schwefel, H.-P.: Evolutionäre Algorithmen: Ein robustes Optimierkonzept. Physikalische Blätter 50(3), 236–239 (1994)

Rushmeier, R.A., Kontogiorgis, S.A.: Advances in the optimization of airline fleet assignment. Transportation Science 31(2), 159–169 (1997)

Russon, M.G., Riley, N.F.: Airport substitution in a short haul model of air transportation. International Journal of Transportation Economics XX(2), 157–173 (1993)

Ryan, D.M.: The solution of massive generalized set partitioning problems in air crew rostering. Journal of the Operational Research Society 43, 459–467 (1992)

Sarac, A., Batta, R., Rump, C.M.: A branch-and-price approach for operational aircraft maintenance routing. European Journal of Operational Research 175, 1850–1869 (2006)

Sarma, J., De Jong, K.A.: Generation gap methods. In: Bäck, T., Fogel, D.B., Michalewicz, Z. (eds.) Handbook of Evolutionary Computation, pp. C2.7:1–C2.7:5. Institute of Physics Publishing. Oxford University Press, Oxford (1997)

Schmitz, G., Kauthen, J.-P., Akkaya, N., Halsall, L.: NAPPA mini model guide. User Information. Zurich (1998)

Seristö, H., Vepsalainen, A.P.J.: Airline cost drivers: cost implications of fleet, routes, and personnel policies. Journal of Air Transport Management 3(1), 11–22 (1997)

Shen, G.: Reverse-fitting the gravity model to inter-city airline passenger flows by an algebraic simplification. Journal of Transport Geography 12, 219–234 (2004)

Sherali, H.D., Bish, E.K., Zhu, X.: Airline fleet assignment concepts, models, and algorithms. European Journal of Operational Research 172, 1–30 (2006)

Skytrax. World airline star ranking (2006) (May 22, 2006), http://www.airlinequality.com/StarRanking/ranking.htm

Smith, B.C., Günther, D.P., Rao, B.V., Ratliff, R.M.: E-commerce and operations research in airline planning, marketing, and distribution. Interfaces 31(2), 37–55 (2001)

Smith, B.C., Leimkuhler, J.F., Darrow, R.M.: Yield management at American Airlines. Interfaces 22(1), 8–31 (1992)

Sohoni, M.G., Johnson, E.L., Bailey, T.G.: Long-range reserve crew manpower planning. Management Science 50(6), 724–739 (2004)

Solomon, M.M., Desrosiers, J.: Time window constrained routing and scheduling problems. Transportation Science 22(1) (1988)

Song, M., Wei, G., Yu, G.: A decision support framework for crew management during airline irregular operations. In: Yu, G. (ed.) Operations Research in the Airline Industry, pp. 259–286. Kluwer Academic Publishers, Boston (1998)

Soumis, F., Ferland, J.A., Rousseau, J.-M.: A model for large-scale aircraft routing and scheduling problems. Transportation Research Part B: Methodological 14, 191–201 (1980)

Sriram, C., Haghani, A.: An optimization model for aircraft maintenance scheduling and re-assignment. Transportation Research Part A: Policy and Practice 37(1), 29–48 (2003)

Stojkovic, M., Soumis, F.: An optimization model for the simultaneous operational flight and pilot scheduling problem. Management Science 47(9), 1290–1305 (2001)

Stojkovic, M., Soumis, F., Desrosiers, J.: The operational airline crew scheduling problem. Transportation Science 32(3), 232–245 (1998)

Storer, R.H., Wu, S.D., Vaccari, R.: Problem and heuristic space search strategies for job shop scheduling. ORSA Journal of Computing 7(4), 453–467 (1995)

Su, Y.Y., Srihari, K.: A knowledge based aircraft-gate assignment advisor. Computers and Industrial Engineering 25(1-4), 123–126 (1993)

Subramanian, J., Scheff Jr., R.P., Lautenbacher, C.J.: Airline yield management with overbooking, cancellations, and no-shows. Transportation Science 33(2), 147–167 (1999)

Subramanian, R., Scheff Jr., R.P., Quillinan, J.D., Wiper, D.S., Marsten, R.E.: Coldstart: Fleet assignment at Delta Airlines. Interfaces 24(1), 104–120 (1994)

Suhl, L.: Entscheidungsunterstützende Systeme bei der Produktionsplanung in Fluggesellschaften. Wirtschaftsinformatik 35(6), 542–550 (1993)

Suhl, L.: Computer-aided scheduling, an airline perspective. Gabler, Wiesbaden (1995)

Suzuki, Y.: The relationship between on-time performance and airline market share: a new approach. Transportation Research Part E: Logistics and Transportation Review 36, 139–154 (2000)

Syswerda, G.: Uniform crossover in genetic algorithms. In: Proceedings of the 3rd International Conference on Genetic Algorithms, pp. 2–9 (1989)

Tacke, G., Schleusener, M.: Bargain airline pricing – how should the majors respond? Travel & Tourism White Paper, SIMON, KUCHER & PARTNERS. Bonn et al (2003)

Taha, H.A.: Operations research: An introduction, 7th edn. Prentice Hall, Englewood Cliffs (2002)

Talluri, K.T.: Swapping applications in a daily airline fleet assignment. Transportation Science 30(3), 237–248 (1996)

Talluri, K.T.: The four-day aircraft maintenance routing problem. Transportation Science 32(1), 43–53 (1998)

Taneja, N.K.: Driving airline business strategies through emerging technology. Ashgate Publishing Company, Hants (2002)

Teodorovic, D.: A model for designing the meteorological most reliable airline schedule. European Journal of Operational Research 21, 156–164 (1985)

Teodorovic, D.: Airline operations research. Transportation Studies. Gordon and Breach, New York (1988)

Teodorovic, D.: Model to reduce airline schedule disturbances. Journal of Transportation Engineering 121(4) (1995)

Teodorovic, D., Kalic, M., Pavkovic, G.: The potential for using fuzzy set theory in airline network design. Transportation Research 28(2), 103–121 (1994)

Teodorovic, D., Krcmar-Nozic, E.: Multicriteria model to determine flight frequencies on an airline network under competitive conditions. Transportation Science 23(1), 14–25 (1989)

Thengvall, B.G., Bard, J.F., Yu, G.: Balancing user preferences for aircraft schedule recovery during irregular operations. IIE Transactions 32(3), 181–193 (2000)

Thengvall, B.G., Bard, J.F., Yu, G.: A bundle algorithm approach for the aircraft schedule recovery problem during hub closures. Transportation Science 37(4), 392–407 (2003)

Thengvall, B.G., Yu, G., Bard, J.F.: Multiple fleet aircraft schedule recovery following hub closures. Transportation Research Part A: Policy and Practice 35, 289–308 (2001)

Train, K.E.: Discrete choice methods with simulation. Cambridge University Press, Cambridge (2003)

Trietsch, D.: Scheduling flights at hub airports. Transportation Research Part B: Methodological 27(2), 133–150 (1993)

Ubøe, J.: Aggregation of gravity models for journeys to work. Environment and Planning A 36(4), 715–729 (2004)

Urban, D.: Logit-analyse – statistische verfahren zur analyse von modellen mit qualitativen response-variablen. G. Fischer, Stuttgart (1993)

Vance, P.H., Barnhart, C., Johnson, E.L., Nemhauser, G.L.: Airline crew scheduling: A new formulation and decomposition algorithm. Operations Research 45(2), 188–200 (1997)

Vanderstraeten, G., Bergeron, M.: Automatic assignment of aircraft to gates at a terminal. Computers and Industrial Engineering 14(1), 15–25 (1988)

Verleger Jr., P.K.: Model for the demand for air transportation. The Bell Journal of Economics and Management Science 3(2), 437–457 (1972)

Verwijmeren, M.A.A.P., Tilanus, C.B.: Network planning for scheduling operations in air cargo handling: a tool in medium term goodsflow control. European Journal of Operational Research 70, 159–166 (1993)

Vinod, B.: Origin and destination yield management. In: Jenkins, D. (ed.) The Handbook of Airline Economics, pp. 459–468. The Aviation Weekly Group of the McGraw-Hill Companies, New York (1995)

Weatherford, L.R.: A tutorial on optimization in the context of perishable-asset revenue management problems for the airline industry. In: Yu, G. (ed.) Operations Research in the Airline Industry, pp. 68–100. Kluwer Academic Publishers, Boston (1998)

Weatherford, L.R., Gentry, T.W., Wilamowski, B.: Neural network forecasting for airlines: A comparative analysis. Journal of Revenue and Pricing Management 1(4), 319–331 (2003)

Wei, G., Yu, G., Song, M.: Optimization model and algorithm for crew management during airline irregular operations. Journal of Combinatorial Optimization 1, 305–321 (1997)

Wei, W., Hansen, M.: Impact of aircraft size and seat availability on airlines' demand and market share in duopoly markets. Transportation Research Part E: Logistics and Transportation Review 41, 315–327 (2005)

Wei, W., Hansen, M.: Airlines' competition in aircraft size and service frequency in duopoly markets. Transportation Research Part E: Logistics and Transportation Review (2006)

Wells, A.T.: Air transportation: A management perspective, 4th edn. Wadsworth, Belmont (1998)

Wendt, O.: Tourenplanung durch einsatz naturanaloger verfahren. integration von genetischen algorithmen und simulated annealing. Gabler, Wiesbaden (1995)

Whitley, D., Kauth, J.: Genitor: A different genetic algorithm. In: Rocky Mountain Conference on Artificial Intelligence, pp. 118–130 (1988)

Wojahn, O.W.: The impact of passengers preferences regarding time and service quality on airline network structure. Journal of Transport Economics and Policy 36(1), 139–162 (2002)

Wolpert, D.H., Macready, W.G.: No free lunch theorems for optimization. IEEE Transactions on Evolutionary Computation 1(1), 67–82 (1997)

Woodside, J.: ICAO world airfield catalogue (December 10, 2006),
http://www.homepages.mcb.net/bones/06airfields/icao.htm

Wu, C.-L.: Inherent delays and operational reliability of airline schedules. Journal of Air Transport Management 11(4), 273–282 (2005)

Wu, C.-L., Caves, R.E.: Aircraft operational costs and turnaround efficiency at airports. Journal of Air Transport Management 6(4), 201–208 (2000)

Wu, C.-L., Caves, R.E.: Towards the optimisation of the schedule reliability of aircraft rotations. Journal of Air Transport Management 8(6), 419–426 (2002)

Yan, S., Chang, C.-M.: A network model for gate assignment. Journal of Advanced Transportation 32(2), 176–189 (1998)

Yan, S., Chang, J.-C.: Airline cockpit crew scheduling. European Journal of Operational Research 136(3), 501–511 (2002)

Yan, S., Huo, C.-M.: Optimization of multiple objective gate assignments. Transportation Research Part A: Policy and Practice 35, 413–432 (2001)

Yan, S., Tang, C.-H.: A heuristic approach for airport gate assignments for stochastic flight delays. European Journal of Operational Research 180(2), 547–567 (2007)

Yan, S., Tseng, C.-H.: A passenger demand model for airline flight scheduling and fleet routing. Computers and Operations Research 29(11), 1559–1581 (2002)

Yan, S., Tu, Y.-P.: Multifleet routing and multistop flight scheduling for schedule perturbation. European Journal of Operational Research 103(1), 155–169 (1997)

Yan, S., Tu, Y.-P.: A network model for airline cabin crew scheduling. European Journal of Operational Research 140(3), 531–540 (2002)

Yan, S., Tung, T.-T., Tu, Y.-P.: Optimal construction of airline individual crew pairings. Computers and Operations Research 29(4), 341–363 (2002)

Yau, C.: Dynamic flight scheduling. OMEGA International Journal of Management Science 17(6), 533–542 (1989)

Yen, J.W., Birge, J.R.: A stochastic programming approach to the airline crew scheduling problem. Transportation Science 40(1), 3–14 (2006)

Yu, G.: Operations research in the airline industry. Kluwer Academic Publishers, Boston (1998)

Yu, G., Argüello, M.F., Song, G., McCowan, S.M., White, A.: A new era for crew recovery at Continental Airlines. Interfaces 33(1), 5–22 (2003)

Zeghal, F.M., Minoux, M.: Modeling and solving a crew assignment problem in air transportation. European Journal of Operational Research 175, 187–209 (2006)

Glossary

airline schedule flight schedule including crew and fleet assignment and routing/rotation information

block time time period between departure from the gate (*off blocks*) at the departure airport until arrival (*on blocks*) at destination airport (block time = flight time + taxi time)

connection sequence of two or more individual flights

crew pairing multi-day sequence of flight legs flown by the same crew

fleet type representative example of all aircraft with the same operating characteristics

flight schedule set of flights with departure and arrival time and airport information (*timetable*)

flight time duration of a flight from take-off to landing

ground time time an aircraft is idle on ground

hub airport with many connection possibilities

hub-and-spoke network (flight) network with non-stop flights from each lower demand airport (spokes) only to one major airport (hub)

itinerary travel alternative between two airports (either a nonstop flight or a sequence of connecting flights)

maintenance routing routing that contains maintenance stations for the corresponding fleet type to satisfy maintenance constraints

maintenance station airport at which the maintenance for a specific fleet type can be conducted

market combination of time and origin & destination (O&D) where competition takes place

market size total number of passengers that want to travel by air in a given market

maximum connection time time a passenger is willing to wait for a connecting flight

minimum connection time time necessary for passengers to change to a connecting flight and to process their baggage between the two aircraft

multi-airport city city with more than one airport (for example Berlin: Schönefeld, Tegel, Tempelhof)

origin & destination (O&D) departure and arrival airport of an itinerary

point-to-point network (flight) network with non-stop flights between all airports

ready time point in time at which an aircraft is ready to depart (ready time = arrival time at gate + turn time)

rotation sequence of routings connected to a cycle

route (geographic) link between two or more airports

routing sequence of flights flown by a single aircraft in succession

seat load factor passenger kilometers flown as a percentage of seat kilometers available

tail number distinct registration number assigned to individual aircraft

taxi time time necessary to taxi on ground (to/from gate/runway)

time-of-day curve distribution of passengers' preferred departure (or arrival) times over the day

turn time time that is necessary to prepare an aircraft after landing for the next flight

wave shorter period in time in which many flights arrive or depart at a hub

Printed in the United States
139072LV00001B/84/P